Growing from Cuttings

Growing from Cuttings
and other means

Janet Browne

WARD LOCK LIMITED · LONDON

© Ward Lock Limited 1981

First published in Great Britain in 1981
by Ward Lock Limited, 47 Marylebone Lane,
London W1M 6AX, a Pentos Company.

House editors Denis Ingram and
Deborah Maby

Layout by Charlotte Westbrook

Text phototypeset in Times by Input
Typesetting Ltd, London SW19 8DR

Printed and bound in Great Britain by
Hollen Street Press Ltd, Slough, Berks.

British Library Cataloguing in Publication Data

Browne, Janet
   Growing from cuttings.
   1. Plant propagation
   I. Title
   631.5′3   SB119

   ISBN 0–7063–5993–3 Pbk

Contents

Preface

Increasing plants from your own stocks, or from a piece given to you by a friend or neighbour, is a satisfying experience; it also saves the cost of buying new plants.

There are two main methods by which plants can be increased—from cuttings, the method known as vegetative or asexual propagation, and from seed, known as sexual propagation. The emphasis in this book is on how to propagate plants vegetatively, not only because it is generally the easiest method with many plants but also because it is inexpensive, often does not require sophisticated or expensive equipment, and one does not end up with a surplus of plants, which is often the case when they are raised from seed. Another important factor is that plants propagated vegetatively will be identical to their parents (except possibly when grafting or budding on rootstocks of varying vigour), whereas plants raised from seed harvested at home, or even those sometimes purchased from seedsmen, may not be true to type.

Whatever method is chosen to increase plants vegetatively, it is essential that the parent plant should be healthy and that only healthy pieces are used. Equally, general hygiene of the new plant's growing environment—by disinfection of equipment, sterilized composts, clean water, use of outdoor soil free from pests, diseases and weeds—is important to ensure a strong, long-lived plant.

There is one final point to be remembered when increasing plants. If they are for your own use, or are to be given as a gift to a friend, all well and good. If, however, they are to be sold, care must be taken not to break the laws of the British Plant Varieties and Seeds Act of 1964. This Act enables nurserymen and seedsmen to patent any newly raised plant which is different from any other already in commerce and which maintains its stability of character. The rights are granted for 15 years, and during that period it is an offence for anyone to sell such plants without a licence from the raiser and the payment of a fixed royalty per plant to him. Most catalogues indicate which are patented plants but, if in doubt and the intention is to sell the plants, check the position beforehand.

J.B.

Acknowledgements

The publishers gratefully acknowledge the following persons, agencies and company in granting permission to reproduce the following colour photographs: Pat Brindley (p. 99, lower); R. J. Corbin (p. 36, lower); Humex Ltd. (p. 36, top); Leslie Johns (p. 89, lower); Harry Smith Horticultural Photographic Collection (pp. 72, 89, top, and 90); and Michael Warren (pp. 17, 99, top and 100).
The photograph on p. 18 is by Robert Challinor.
The following line drawings are by Nils Solberg: Figs. 1, 2, 3, 4, 8, 12b, 14b, and 16. Fig. 1 is after the illustration on p. 20 in *The Humex Book of Propagation*, J. Harris, published by Macdonald, 1980; Figs. 8, 12b and 14b are after illustrations on pp. 779, 775 and 780 respectively in *Reader's Digest Encyclopaedia of Garden Plants and Flowers*, 1978.

1 Equipment for Simple Propagation

For the vegetative propagation of hardy plants, the equipment required is minimal over and above that used for the normal cultivation of a garden. Perhaps the most essential requirement is the provision of a warm, sheltered area of the garden, preferably facing south or south-west, and which is protected from cold winds. In addition, the soil should be in 'good heart', that is, it should be reasonably free-draining, easily workable, and contain sufficient organic matter so that air and moisture are retained, and have adequate supplies of nutrients for the growth of young plants.

Such an area is usually called a nursery-bed and it is advisable to keep it for this purpose at all times, rather than to set out cuttings (or sow seeds, except hardy annuals) indiscriminately wherever there is a spare patch of ground. It is well worth caring for the nursery-bed at all times, forking it over, weeding and adding fertilizers as and when necessary. It is, of course, quite possible to use either the nursery-bed, or keep aside a part of it, as a seed-bed.

GENERAL EQUIPMENT

The most obvious hand tools that will be required for propagating plants are spades and forks, both the light border type and the heavier digging sort, a rake—or preferably two, one fine-toothed for creating a crumbly tilth and a larger toothed one for breaking down clods of earth—and two hoes, a Dutch one, which slides backwards and forwards through the soil easily, and a draw-hoe, which has a curved neck and is useful for making drills for cuttings or seeds, as well as chopping out weeds. Smaller tools include secateurs, a sharp gardening knife (or razor blades), a budding knife if possible, a hand trowel, a hand fork, a dibber, preferably about 2.5 cm (1 in) in diameter, (though a pencil will often serve the purpose), a garden line (a strong and long piece of string with stakes at either end), and a watering can, preferably with interchangeable roses so that fine or coarse sprays of water can be applied, or a hose with an adjustable spray nozzle.

In addition, other useful items of equipment are a good supply of labels (both the tie-on type and the sort to be stuck in the soil), wooden pegs or pieces of bent wire (or hairpins) for holding down shoots or stems in the soil, raffia, twine or proprietary ties for budding and grafting, also grafting wax, bags and sheets of polythene, and a variety of canes or other supports. Finally, glass or plastic cloches can be invaluable for forcing early and late crops or for giving protection to young plants until they are able to withstand all weather conditions.

POTS AND BOXES
These are required for many forms of propagation and it is wise to keep to hand a good supply of them, in a variety of sizes, in a sound and clean condition (Fig. 1). In general, pots are made of clay or plastic; it is easier with the former to tell whether or not they need watering and, being porous, they promote rooting, but the latter are more readily cleaned, stored and are usually less liable to break. There are also 'trays' of plastic pots in small sizes. Other types of pots include individual, or strips, of peat pots or blocks, some of which pack very flat but when soaked swell up, the peat being held in position by a fine nylon mesh. Also available on the market are block-making tools for making one's own pots of a special peat/soil mixture. The advantage of such pots and blocks is the lack of root disturbance when the young plants are moved to their final growing positions.

Boxes are sometimes of wood or, more often these days, of plastic. They are available in a variety of sizes. Plastic trays have the advantage of being easier and cleaner to handle and less likely to harbour pests and diseases.

9

a

b

c

d

e

f

g

h

i

j

It used to be considered necessary to crock—place pieces of broken pots in the base of pots and boxes—to assist drainage of the growing medium. This has now been found to be generally unnecessary and, unless there is a blockage of the drainage holes at the bottom, it is only necessary to fill pots and boxes with the required compost. Another advantage of not crocking is that if basal capillary-type watering is adopted (see Chapter 2), the compost in the container can draw up water more evenly and remain at the required state of moisture.

COMPOSTS AND FERTILIZERS
When propagating hardy plants in the open, it is usual to insert the young plants, cuttings or seeds directly into ground that has been well prepared in advance. As mentioned earlier, the soil should be reasonably light and workable and ideally, coarse garden sand and/or peat should be worked into it. These two materials promote quicker rooting by helping to ensure that the soil is reasonably free-draining yet moisture-retentive, thus creating an environment conducive to healthy growth.

When plants are to be increased under what might be termed the artificial conditions of a greenhouse, frame or in the home, it is inadvisable to use soil brought in from the garden. It may carry weed seeds, pests and diseases and, in the generally warm, moist conditions created for propagation, these can be a considerable problem. Equally, the soil itself may well not be of suitable quality.

For these reasons it is strongly advised that suitable proprietary composts are purchased. There are basically two types available—those based on good sterilized soil and those which contain ingredients other than soil and which are referred to as soilless composts. The most commonly available soil-based composts are John Innes composts, to which sand, peat and fertilizers have been added in different

Fig. 1 Pots and boxes for propagation. (a) Clay pots. (b) Standard seed tray with ventilated plastic dome. (c) Polystyrene slab pots. (d) Fibre pots. (e) Standard and half-sized boxes. (f) Strips of peat pots. (g) Round and square peat pots. (h) and (i) Jiffy 7 compressed peat pots. (j) Plastic pots.

11

ratios according to their strengths. The mildest of these is the John Innes seed compost; next in strength is J.I. potting compost no. 1, followed by potting composts nos. 2, 3 and 4, the last of these being the strongest and generally used only for large, container-grown houseplants. There is also a John Innes cuttings compost.

Soilless composts also come in different strengths, and only the mildest should be used for rooting cuttings or seed sowing. When using soilless composts, care with watering is necessary. They tend to hold water more than soil-based ones, so must not be over watered; equally, if they are allowed to dry out completely, they can be very difficult to re-wet evenly and thoroughly.

Silver sand is a useful commodity to have around, both for helping in the rooting of cuttings and the covering of seeds. Equally useful is a bottle or packet of a general-purpose proprietary fertilizer for giving liquid feeds to young plants when there may be a delay in transplanting and the compost in which they are growing has insufficient plant nutrients to maintain healthy growth. A general-purpose granular type of fertilizer is useful for adding to outdoor nursery- and seed-beds. Apply fertilizers at the rate recommended by the manufacturers.

PESTICIDES

The most common problem affecting young plants and seedlings is damping-off disease. Watering with Cheshunt Compound helps prevent this. Other useful pesticides to keep to hand are those for controlling red spider mite, aphids (green- and black-fly), white fly, woodlice, earwigs, and powdery and downy mildews.

2 Equipment for Raising Plants in Heat

As discussed in later chapters, heat is required in order to propagate certain plants. The amount of heat can vary considerably, from a mild 7°C (45°F) for general growth of many plants to a tropical temperature of 24°C (75°F) or more for tender greenhouse and houseplants. In order to provide these heat requirements, and the necessary protection from outdoor weather conditions, a special unit such as a greenhouse or frame, or certain types of equipment, namely propagating units, are usually required. On a small scale however, it is quite possible to increase plants in warm rooms in the house with a little ingenuity and improvisation.

GREENHOUSE
There are many types of greenhouses available, both free-standing and lean-to's, made of a variety of materials. Whichever type is purchased, site it in a sheltered position where it gets plenty of light—this being particularly important as many plants are propagated during the winter and spring months when light intensity is low. It can always be shaded with blinds or special paint in the summer if necessary. Ensure there is adequate bench space and enough shelves for your requirements—or facilities for fixing additional ones—and an area of open bed if it is intended to grow greenhouse plants at soil level. It is also important in the average-sized garden greenhouse to have a central path, preferably of concrete as this is easy to clean.

A source of heat in the greenhouse will be essential for propagating plants. A unit without any heat at all, referred

13

to as a cold house, will be useless for raising plants because during the winter months the temperature inside will be as low, if not lower, than that outside.

Heating

There are several ways in which a greenhouse can be heated—by using electricity, gas, paraffin, or oil or solid fuel boilers—but for ease of maintenance, and the added advantage of being able to power other items of greenhouse equipment, electricity is hard to beat. A secondary source of heating, such as a paraffin stove or bottled propane gas unit, is a useful precaution against power cuts. This can also be used to heat up one corner of the greenhouse (which can be screened off with polythene) to a higher temperature—here more heat may be provided for propagating certain plants and maintaining mature hot-house subjects than is supplied to the house as a whole.

Whatever the source of power for heating, running costs continually increase and it is obviously advantageous financially to keep them down as low as possible. Insulating the greenhouse with a form of plastic sheeting, particularly the bubble type (which is two layers of polythene welded together with air bubbles trapped between), keeping the vents closed, and making sure the door and structure of the greenhouse are draught-proof, will all help. So too, will a small electric fan installed in the top corner of the greenhouse, opposite the door, and placed facing downwards at an angle of 10°. It is inexpensive to operate and its function is to drive back down to plant level the hot air that rises and collects in the roof space. As it is unlikely that the whole greenhouse will be used for propagation purposes at any one time a more highly heated greenhouse within a greenhouse can be created with polythene sheets, as previously suggested.

Another method of conserving heat and reducing costs is the use of electric soil-warming cables on the bench. Plants, cuttings and seeds grow, root, and germinate best when the temperature in the root area is higher than that around the shoots, stems and leaves. The cables are sandwiched in the middle of a 7.5–10 cm (3–4 in) layer of coarse, sharp sand

14

(Fig. 2), and are run up and down the bench in such a way that at no point do they touch (follow the manufacturer's instructions carefully). To ensure the required even temperature, instal a rod-type thermostat. As the sand needs to be damp in order to conduct heat, a sheet of automatically water-fed capillary matting can be laid below the sand; alternatively, the sand can be kept moist by watering with a can. Pots or boxes of cuttings, and plants in pots, are then stood on this damp sand, and moist peat is used to fill the spaces between them to spread the heat round the root areas. As the weight to be carried by the bench will be heavy, make sure it is soundly constructed. Also, to keep the sand and peat in position, surround it with 20–23 cm (8–9 in) high boards and, if the bench is slatted, cover it first with asbestos sheeting, roofing felt, timber or similar material to give a solid, flat surface.

By keeping the temperature around the roots at 13°–16°C (55°–61°F) using soil-warming cables, a minimum air temperature of 7°C (45°F) at night will be quite sufficient for

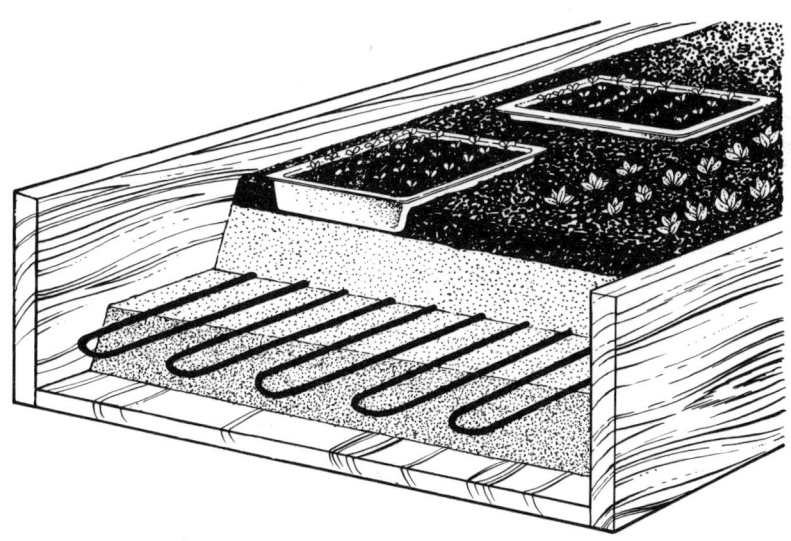

Fig. 2 Part of the greenhouse bench converted to raise new plants. A 'box' is constructed, a layer of damp sand put in the base, a soil warming cable installed, and moist peat placed around boxes and pots.

most plants. This will represent a considerable saving in fuel costs as, whatever fuel is used, for every degree rise in temperature the price will start going up, to such an extent that at 10°C (50°F) it will be double that at 7°C (45°F). On the other hand, the cost of operating soil-warming cables is minimal in comparison.

Once soil-warming cables have been installed, it is a simple matter to turn a part of the bench into a propagating case for cuttings. Merely create a tent structure of glass or polythene, or even use a cloche, so that the cuttings are completely enclosed in their own warm and moist environment. Alternatively, just cover the pots and boxes with glass or polythene as recommended in later chapters.

Water

A supply of fresh water laid on to the greenhouse may sound extravagant but it is well worth considering for three main reasons. Firstly, it is extremely labour-saving to have water to hand and not to have to carry cans or use a hose. Secondly, using clean fresh water does much to prevent introducing possible pests and diseases from rain water barrels, which can be a bad source of trouble unless frequently cleaned and disinfected. And thirdly, it enables use to be made of modern watering techniques which do much to assist healthy plant growth.

As mentioned in the previous section under 'Heating', the use of capillary matting, which is automatically fed with water, in conjunction with soil-warming cables, will allow the compost in the pots and boxes to absorb, by capillary action, the amount of water they require to keep moist. A secondary advantage is that the atmosphere round the plants is kept slightly damp. A small 'wick' of fibreglass or similar material may need to be inserted into the compost through the basal hole of each pot, leaving the other end free in the sand, to ensure adequate water absorption.

There are basically two methods by which the water can be fed automatically to greenhouse benches covered with capillary matting or, if heating cables are not used, just plain sand or grit chippings. Both involve the use of a water jar or tank, which is fed direct from the main supply by a

To prevent evaporation of moisture from the leaves of cuttings when taken in quantity, place them in a polythene bag until ready for preparing for planting.

A well-arranged greenhouse with an electrically operated propagator for raising new plants from cuttings or seeds and plenty of floor, bench and shelf space for standing pots and boxes. It is heated by a paraffin stove.

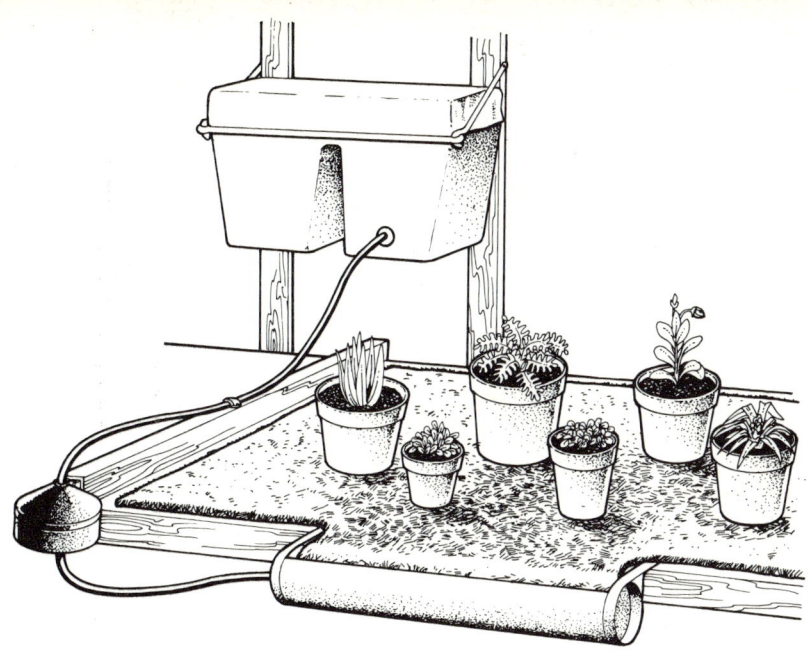

Fig. 3 Automatic bench watering. A capillary mat on the bench is kept moist by a constant level of water drawn up from the trough which is kept filled by the valve which controls the water flow from the tank above. The pot plants draw up the water from the wet matting by capillary action.

hose or a watering can, and a special, sometimes adjustable valve, which releases water at a specified rate. The water is distributed to the plants standing on the sand, grit, or capillary matting, either by the action of capillary wicks or the matting itself, which has one side dipped into the front-of-bench trough into which water has been released (Fig. 3). Alternatively, the water is distributed through trickle irrigation lines which have drip nozzles, so that pots or boxes can either be watered individually, or the bench covering material as a whole.

Such automatic systems are ideal for maintaining the growing medium at the correct stage of dampness for rooting cuttings and increasing plants generally, as well as maintaining mature specimens with the minimum of effort. They also supply plants with water the natural way—via the roots—and require little or no maintenance. Liquid fertilizers can be added to the water if required, and a special algicide applied to the sand, grit or capillary matting will prevent the growth of algae and inhibit the plant roots from growing through the base of their pots or boxes. As with heating equipment, these types of watering units are sold in kit form and there are a number on the market. Whichever is selected, follow the manufacturer's instructions closely; if in doubt about the installation of electrical equipment or mains water supplies, always seek the guidance of an expert.

If the more sophisticated types of watering equipment are not to be installed in your greenhouse, keep a watering can with a fine rose specially for watering plants that are being propagated. Also a sprayer, so that fine sprinklings or sprays of clean water can be given. Try not to give cold tap water as this can check young plants; leave the full can or sprayer in the greenhouse for a few hours so that it can warm up a little before being applied. This same rule applies to the watering of the sand on the benches, if done by can.

Mist Propagating Unit
This is essentially the bringing together of bench-warming and automatic watering equipment to provide the ideal conditions for raising plants from cuttings.

A ready-made mist propagating unit kit can be purchased complete, or the bench part can be made up as described under the 'Heating' section, with the addition of panes of glass stood adjacent to each other inside the wooden bench surround to give a close sided case standing some 75 cm (2 ft 6 in) high. The mist unit consists of an electro-mechanical detector which operates according to the environment in the unit; it can be pre-set according to requirements. When there is a danger of the young cuttings drying out, the detector pad also dries out and its change in weight tips a balance arm which, in turn, triggers a switch to give a fine spray of water. As soon as the detector pad and cuttings are moist enough, the weight tips the balance arm the opposite way, and so turns off the water. In this manner, the cuttings get the moisture they require, but the rooting medium never gets waterlogged. Some mist propagating units have a 'weaning' device, so that the rooted cuttings can be gradually hardened off to become acclimatized to normal growing conditions before being transplanted.

Propagating Cases

Electrically heated proprietary propagating units are invaluable for increasing plants. They come in a variety of shapes and sizes, are cheap and economical to operate, and can be used just as well in the home as in the greenhouse. They are usually constructed of fibreglass or some form of rigid plastic with built-in heating wires and thermostat in the base. They are supplied with clear plastic dome tops, sometimes in a variety of shapes and sizes to accommodate different heights and spreads of pots, boxes and the plants themselves. The more sophisticated models have two pilot lights, one to show when the mains electrical supply is on and another to show when the thermostat is working.

To use an electrical propagating case, place it on a flat surface, put a layer of moist coarse sand in the base, and stand the prepared pots or boxes of cuttings or seeds directly on this. Put the top over the case and, if necessary to provide darkness or shade, drape paper or material over it.

The range of temperatures provided by such units is roughly from 7° to 30° C (45° to 85° F), which means a very wide range of plants can be propagated. And where the unit has a tall covering dome, it is an economical 'home' for tropical plants which require high growing, or what are sometimes called, stove temperatures. An added refinement to some propagating cases is the incorporation of adjustable air vents, to help in the hardening off process of rooted cuttings and seedlings, before the dome is removed completely.

FRAMES

Frames, either heated or unheated, play a large part in the process of increasing plants in the garden. If heated, they can often take the place of a greenhouse when only a few plants are to be propagated at any one time. If unheated, they are the ideal place for hardening off plants between the warmth of a greenhouse and the cold of outdoors, or for giving that little bit of weather protection that so often helps cuttings to root and seeds to germinate more quickly and reliably than they would if exposed to the vagaries of the climatic elements.

Frames are available in a variety of sizes and are made of a number of different materials. They should have a sloping glass or plastic light (lid) which can be slid easily up and down for ventilation purposes, or be removed entirely at the end of the hardening off period.

A frame should be sited in a sunny, sheltered position. If it is to be heated, it is best done electrically, either by running cables or tubular heaters round the walls, or by using soil warming cables, on the same principle as described for bench heating in the greenhouse, or a combination of both forms of heating. The heating should be thermostatically controlled to give the required temperatures and, to help reduce heating costs in cold weather, old blankets, rugs, curtains or sacking can be placed over the top at night to help maintain a warmer interior.

Although cuttings can be inserted, or seeds sown, directly in a well-prepared compost in a heated or cold frame, it is far preferable to use pots and boxes, which can be handled easily, removed as soon as possible to give space for others, and which prevent pests and diseases entering the soil unnecessarily.

The general management of frames for propagating plants is the same as that described previously for greenhouses, with the exception that it is not so easy to install automatic watering devices, so this operation will generally need to be carried out by hand.

RAISING PLANTS IN THE HOME

The lack of a greenhouse or frame is not a disaster if you want to propagate plants, for many can well be raised in the home in a warm room which gets plenty of light.

An electrically operated propagating case, such as described in the greenhouse section of this chapter, can easily be used indoors and the plants inside maintained as they would be if the case was in the greenhouse. Once rooted, the young plants can be hardened off by placing them on window sills then, if they are to be grown on outdoors, they can be moved to cooler, then cold rooms, until finally they can be placed under cloches in a sheltered place outdoors prior to planting out.

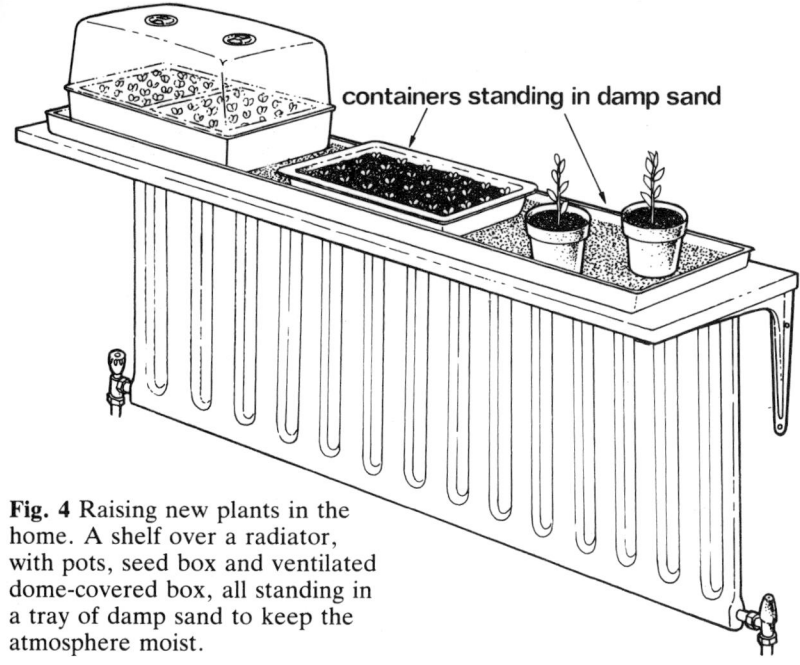

containers standing in damp sand

Fig. 4 Raising new plants in the home. A shelf over a radiator, with pots, seed box and ventilated dome-covered box, all standing in a tray of damp sand to keep the atmosphere moist.

If you have no special propagating equipment, the prepared pots and boxes of cuttings and seeds can be placed in transparent polythene bags with their tops tied firmly to prevent the entry of air. Use twigs or canes to hold the plastic away from the leaves of the plants. Keep the plants wrapped like this in a warm place until growth takes place, then gradually increase ventilation by opening the bags until the plants are strong enough to grow without them and are ready to be transplanted or potted up. If the bagged pots and boxes, or new young uncovered plants in their containers, are placed on window sills (preferably stood on plastic trays filled with a layer of moist sand or grit to give a humid atmosphere) make sure they don't suffer from cold draughts from the window.

Standing containers of cuttings or seeds either on radiators or shelves above (Fig. 4) can be effective provided the radiators are not too hot and the containers are stood in trays of sand that is kept regularly moist, to give humidity to overcome the dry atmosphere caused by central heating. Syringing the young plants regularly with water will be necessary in any home which is centrally heated.

3 Simple Plant Division

The easiest method of increasing hardy plants (particularly herbaceous perennials, alpines and a few shrubs, fruits and vegetables) is by division. This can be carried out in various ways and it involves dividing clumps of plants into smaller pieces, each carrying several roots and shoots. The best periods to do this are usually in autumn or spring, or any time between the two when weather and soil conditions are suitable. As a general guide, spring-flowering plants are best divided in autumn, and late summer- and autumn-flowering ones in spring. At whatever time division is carried out, put the new plants into the ground again as soon as possible, or into pots or boxes of moist soil if there is to be a delay before planting.

Many herbaceous plants and alpines need dividing regularly every three to five years to prevent the clumps getting too large or dying away in the centre. In such cases it is necessary to lift the plants completely, divide them, discard the older inner section and replant the young outer growths and roots firmly in soil that has been freshly dug and prepared with humus-forming materials such as peat, shredded bark or well-rotted compost with the addition of bone meal or a good general-purpose fertilizer. Alpine plants may additionally require coarse sand or chippings to give them the free draining conditions they require.

Where large areas of plants are to be divided, it is often easier to lift all the plants at once, lay them on a hard path surface or sheets of polythene and cover the roots with damp sacking; this will prevent drying out and wilting. Then the whole border or rock garden soil can be forked over, weeds removed and the humus matter and fertilizers added in one main operation. Tread the ground firm then divide the plants and set out the young portions in the required positions. Do not carry out this type of large operation during hot dry weather or the plants may suffer. Also remember what height and spread the plants achieve, and their flower colours and flowering periods, so that they are planted in the desired positions.

Tender plants, that spend their lives in the home or greenhouse and which can be increased by division, should be divided and replanted in pots or containers in spring.

There are several ways of dividing plants:

BY HAND
Teasing apart roots and shoots of plants by hand is the simplest, easiest and generally most satisfactory method of division. Lift the plant slightly from the soil so that the young pieces, complete with roots and shoots, can be carefully separated from each other with the minimum of damage to them and the parent plant. Then firm back the parent and set out the off-spring.

WITH A KNIFE
If plants are too tough to separate by hand, a sharp knife will often help to cut sections apart in the required manner, still taking care to ensure each new young plant is composed of roots and shoots. A knife is also used to cut the stem between the parent plant and young plant that has formed as a result of natural layering by runners.

USING A SPADE
Clumps of spreading plants, such as low growing and carpeting ones used for ground cover purposes which form new roots from their prostrate stems, can be divided simply by slicing off outside pieces with a sharp-bladed spade. Lift the pieces, with some soil if possible, to avoid root disturbance, and plant where required. Replace soil around the parent plants if necessary.

DIVISION WITH FORKS

Some plants, herbaceous perennials in particular, form large tough clumps and the only way to deal with these satisfactorily is to lift them out of the ground completely and break them apart by inserting the prongs of two forks, placed back to back in the centre of the clump. Lever the forks backwards and forwards until the clump breaks apart, then pull off the young outer pieces by hand. The centre piece which will almost certainly be past giving a good performance is best discarded and the outer young portions retained instead.

AFTER CARE

All young plants obtained by simple division are treated as were their parents. They are planted firmly in well-prepared soil, or potting compost if in pots, watered in to help settle the soil around the roots, and labelled if necessary; they should quickly become well established. During very dry periods, it is advisable to see that the young plants are kept well watered until they are growing strongly, and can fend for themselves. Mulching around the plants will also help in this respect.

4 Division of 'Bulbous' Plants

Plants which have thickened roots or shoots are often all referred to as 'bulbs'. Botanically this is incorrect, for these swollen food storage organs should be divided into four classifications—bulbs, corms, rhizomes and tubers. For the purpose of increasing stock of these various plants it helps to know something about their structure in order to propagate them correctly.

BULBS

A bulb, which may be hardy, half-hardy or tender, is usually a slightly egg-shaped unit with a pointed top and a small flat base; tulips and hyacinths are typical examples. This underground, plump, food-storage organ is really a very modified 'bud' with scaly or fleshy food-supplying 'leaves' growing up round it from the flat basal 'plate'. When conditions are suitable and the bulb is growing, the 'bud' produces stems, true leaves and flowers which grow upwards, and the basal 'plate' puts out roots which grow down into the soil. These latter anchor the bulb and draw up nutrients and water to help replenish those used up by the top growth, which draws from the food supplies in the fleshy bulb 'leaves'. The true above-ground leaves of a bulbous plant also manufacture foods and pass these back into the bulb. Thus the growing strength of the bulb is built up again and a new rudimentary 'bud' is produced to perform well the following season.

Sometimes a bulb may have a small bulb or bulbs attached to it, but these rarely produce flowers until they have built up sufficient food supplies and enlarged in size. But these

small off-sets, or 'daughter bulbs' as they are sometimes called, are what are used to increase a stock of bulbs. Sometimes the parent bulb also divides itself into two or three bulbs, and again these can be used similarly.

The best time to split up small bulbs and parent bulbs is during their dormant season. Simply lift the bulb from the ground, if it is a hardy outdoor type, or empty out the pot of compost and bulbs if of a tender variety, and remove the youngsters with the fingers. Replant each separately and allow a year or two to pass before expecting a healthy floriferous performance from the young stock.

Some lilies are an exception to this method of propagation. As the bulb 'leaf' scales are of a looser formation than the tight ones of most bulbs, each scale can be carefully pulled off and planted shallowly in pots or boxes containing a sandy, peaty compost (Fig. 5). Stand the containers in a cold frame, water as necessary and, after several months a small bulb will form at the base of each scale, usually when new leaves also appear. These are then planted and grown as normal 'daughter bulbs'. They take two to five years to flower. 'Cloves' of *Allium sativum* (garlic) or scales are split from the parent bulb and planted out each spring for harvesting the following autumn.

Hardy outdoor bulbs, such as tulips, narcissi and hyacinths, usually only require dividing if they get overcrowd-

Fig. 5 (*Left*) Propagating lilies and related plants. Carefully remove the 'leaf' scales and plant each in a sandy peaty compost in a pot. New plants will arise from the bases.

Fig. 6 (*Right*) Bulbils for propagation of plants such as some lilies (as shown here) Egyptian onion and some ferns. If removed and planted in boxes or a seed bed, after two or three years they will have grown large enough to be set out where required.

ed and flower poorly, or when additional bulbs are required. The best time to lift and divide snowdrop clumps is immediately after flowering. Tender bulbs, such as lachenalia, freesia and amaryllis, are best divided and repotted annually. Half-hardy bulbs, such as *Allium cepa* (onions), *Allium ascalonicum* (shallots) and *Vallota speciosa* (Scarborough lily) are usually lifted each autumn, stored over winter in a frost-free place, and replanted the following spring.

Bulbils

Some bulbous plants, such as *Lilium tigrinum* (tiger lily) and *Allium cepa aggregatum* (Egyptian onion), produce small rounded bulbils in the axils between the above-ground

29

stems and leaves or instead of flowers (Fig. 6). If these are removed and planted in boxes or an outdoor seed-bed with sandy soil, they will gradually increase in size and, after two or three years, can be planted out where required to perform as normal bulbs. Set them 5–10 cm (2–4 in) apart and deep, according to their size.

CORMS

Corms are not unlike bulbs but are generally more circular in shape, with a pointed 'bud' tip on the upper surface and the root 'plate' below. A corm is primarily made up of a swollen stem base with a few thin scale 'leaves' around it for protection. As with a bulb, the embryo flowers and leaves grow upwards and the roots from the 'plate' downwards and, during growth, utilize the food reserves in the corm. Unlike a bulb, however, new food resources are not fed back into the existing corm but go to form a new one on top of the old one. Typical corms are crocus, hardy, and gladiolus, half-hardy.

As well as producing a new corm each year, small 'cormlets' are also usually formed. If the largest of these are planted normally, they will usually flower the following year. The smaller ones should be planted about 10 cm (4 in) apart and deep in rows in a seed-bed of sandy soil and allowed to grow on for about two years before setting them in their flowering positions. Hardy corms can remain outdoors all year, but those that are half-hardy will require lifting and storing each autumn.

In the case of crocuses, the corms are lifted and the cormlets removed and planted during the dormant season. The division of gladiolus corms can take place when they are lifted in autumn and soil and dead portions removed before quick-drying for winter storage in a frost-free place. Plant out again in late spring.

RHIZOMES

A rhizome, although generally looking like a root, is, in fact, a stem which grows underground or partially above ground. Roots grow from the bottom portion, and leaves, shoots and flowers arise from the top, and young plants are

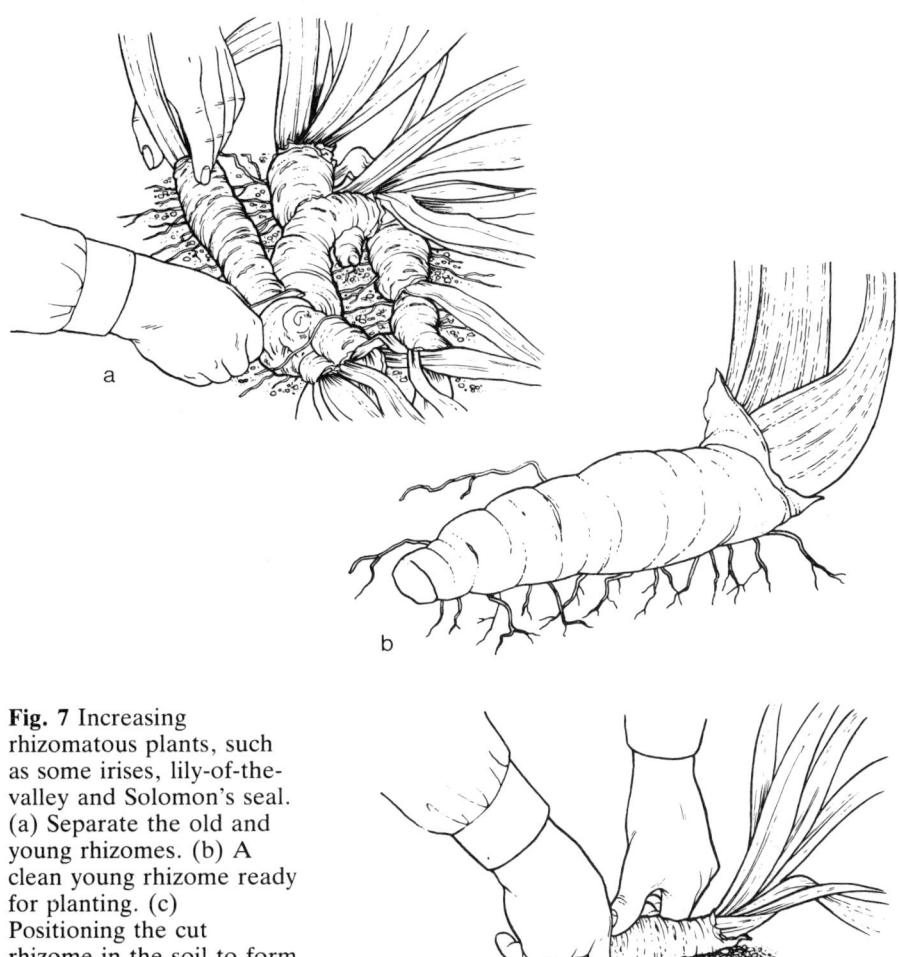

Fig. 7 Increasing rhizomatous plants, such as some irises, lily-of-the-valley and Solomon's seal. (a) Separate the old and young rhizomes. (b) A clean young rhizome ready for planting. (c) Positioning the cut rhizome in the soil to form a new plant.

usually produced at a short distance from the parent plant. A rhizome sometimes looks scaly and jointed and, like bulbs and corms, is a storage organ for nutrients passed to it from the roots and leaves. Typical hardy examples are *Convallaria majalis* (lily-of-the-valley), *Polygonatum × hybridum* (Solomon's seal), flag and German irises (Fig. 7), and the weed couch grass. (This latter should be removed at all cost; it is a menace if allowed to take hold, as each portion of its rhizomes left in the soil forms a new plant.)

31

Propagation of rhizomatous plants is easy, and is best carried out in mid-spring. Dig out the plants carefully, preferably with a fork, and shake off the soil. With a sharp knife, or sharp spade for thick rhizomes, cut each rhizome into pieces, so that each section has some growth buds or shoots above and roots below. Dead or diseased parts are best discarded, but the fresh sections should be replanted immediately, to the same depth in the soil as they were originally. Firm the plants into position and they will quickly start making fresh growth.

TUBERS

Tubers are hardy or half-hardy thickened underground stem or root food-storage organs. As with bulbs, flower stems, shoots and leaves grow upwards and roots downwards, and the food reserves used up during growth are replenished in the tuber by freshly acquired nutrients passed back through the roots and leaves. Tubers vary in shape—they may be hard and rounded or look like very fleshy swollen roots. The difference between a stem tuber and a root tuber is that the former has 'eyes' (buds for new shoot growths) growing on the swollen root-like organ, examples being *Solanum tuberosum* (potatoes), *Helianthus tuberosus* (Jerusalem artichokes), some anemones and tuberous begonias, while a root tuber, such as the dahlia, has 'eyes' only at the base of the aerial stem and not on the tuber itself.

Propagation of hardy perennial tuberous plants can be carried out in spring or autumn. Lift them with a fork, remove the soil, and with a sharp knife cut the tubers into sections, so that each has some eyes and some roots. Replant the new pieces immediately to the same depth as previously, and discard any old or diseased portions. Half-hardy tubers are divided in spring, after they have been lifted in the autumn and stored in a dry, frost-free place during the winter. Stem tubers are cut into sections so that each has one or more 'eyes' which will develop into above-ground growth when planted out. Root tubers require to be cut in such a manner that each piece of 'root' has 'eyes' (growth buds) attached from the aerial stem base, and these stem 'eyes' will form the new top growth.

5 Layering

Layering is an all-embracing term which covers a variety of methods used to increase stocks of plants, but all involve the stems or shoots of the parent plant producing new youngsters, complete with roots and top growth, before they are severed from their parent. Some plants will increase themselves naturally by their stems forming new plants and, unless controlled, can become a nuisance. Others require assistance and need encouragement to provide new stock.

RUNNERS (Sometimes called stolons)
These are shoots that grow above or at ground level and form plantlets, complete with root and top growth, at intervals along their length. Examples of plants that do this naturally are most strawberries, violets and the weed creeping buttercup (which should be eradicated promptly, otherwise it can be a menace).

Even if there is no wish to increase the stock of plants, it is still wise to remove runners, otherwise they tend to form a tangled mass and reduce the vigour of the parent plant.

If it is intended to use the runners for propagation purposes, it is advisable to give nature a helping hand. First of all select only those runners that are vigorous and healthy and cut off all others; a maximum of five runners per plant should be all that are allowed. Then, although they will root naturally, it helps to prepare the soil beneath the rooting points by adding peat or sand. Alternatively, to make later transplanting easier, pots of a peaty, sandy soil mixture can be inserted into the ground (Fig. 8)—sunk into the soil they

Fig. 8 New plants from runners. A strawberry runner planted into a sunken pot of peaty sandy soil where it will form roots. Once established, the young plant's runner is severed from the parent, the pot removed and the youngster set out in its final position.

will retain moisture, though it is still advisable to water during dry conditions. A peg, or even a stone, can be placed over each runner to hold it in position so that it roots where required. To ensure vigorous plantlets, cut off the ends of the runners beyond the selected rooting area to prevent them from forming other plantlets and weakening the selected ones.

Once the new plantlets are growing vigorously, sever the runners between them and the parent plant. Do this several days before transplanting the youngsters, so that they have become completely established on their own and are no longer dependent on food and water supplies from their parent. They will then be ready to be planted wherever required.

Always select runners from healthy plants, especially strawberries, which may well carry debilitating virus diseases that will be transmitted from the parents via the runners to the new plantlets.

The best time to increase plants by means of runners is during late spring and early summer, so they will be ready for planting out in late summer.

SUCKERS

Some plants with roots which spread widely in the soil produce buds from which new shoots and top growth appear. These can be cut with a sharp knife or spade from the parent

34

An electrically heated propagator, large enough to accommodate a tray of cuttings or seedlings. The two adjustable vents at the top allow for ventilation, especially useful when hardening off prior to potting on.

A mist propagator encourages cuttings to root more readily under the controlled conditions of an automatic misting device, which can be adjusted to give the quantity of water required.

Thermostatically controlled soil warming cables in a garden frame will give the required temperatures for rooting many types of cuttings when a greenhouse is not available.

plant, so that each new plantlet is complete with roots and shoots. They can then be planted where required.

This method of plant propagation can be carried out whenever sucker shoots are seen, provided weather and soil conditions are suitable. Examples of plants which may be increased in this manner are *Rubus idaeus* (raspberries), rhus (sumachs) and bamboos.

Some plants which are propagated by grafting (see Chapter 9) or budding (see Chapter 10), may also throw up suckers. Invariably these suckers grow from the stock (root and lower stem portion) of the budded or grafted plant and not from the scion, the top growth, which is the desired plant. As such suckers are usually very vigorous in growth, they may weaken the cultivated plant considerably. Unless they are required for budding or grafting purposes, in which case detach them as described previously and plant them in rows in a nursery-bed, these suckers should be pulled off from as close as possible to the roots or underground stem and destroyed.

SHOOT TIP LAYERING

Plants with long arching stems, such as *Forsythia suspensa*, *Rubus fruticosus* (blackberries) and *Rubus loganobaccus* (loganberries), often root themselves naturally where the shoot tips touch the soil. The new plantlets formed in such a way can be severed from their parent plants and, after a week—to allow them time to recover from the shock of being on their own—be lifted and transplanted to permanent positions. The best time to do this is in early autumn.

To induce tip layering of plants that are not doing so naturally, which includes not only those already mentioned but others such as ribes (currants) and *Grossularia uva-crispa* (gooseberries), bend down pliable but firm young shoots and bury about 10 cm (4 in) of their tips into the soil. (The addition of peat and sand will assist in encouraging quick rooting.) Alternatively, the tips can be buried into pots of a peat, sand, and soil mixture and these sunk into the ground; this reduces transplanting shock later. To keep the tips in position, use pegs or bent wires and keep the layered areas moist by watering during dry weather.

The best time to propagate by tip layering is in summer, so that the new plantlets can be transplanted in the autumn. As with naturally tip layered plants, sever them from their parents a week or so before moving them.

SHOOT LAYERING
With plants that do not layer themselves naturally by runners, suckers or rooting tips, one of the easiest methods of increasing stocks of many trees, shrubs and climbers, also some herbaceous plants, is by encouraging them to root and form plantlets from the stems or shoots of the chosen parent plants. This can be carried out in the open with hardy plants, saves the time, trouble and often expense of propagating by cuttings which require greenhouse or frame conditions, and is a reliable method for producing healthy new stock (provided the chosen parent is healthy in the first place). Not as many new plants can be produced by shoot layering as by cuttings, but for most gardeners, who want only a limited increase in stock, it is definitely a most reliable method, and often succeeds with species which are difficult to propagate by other means.

The best time to layer shoots or stems of trees, shrubs and climbers is either in spring or autumn, though it can often be carried out at any time of year when weather and soil conditions are suitable. Herbaceous plants, such as border carnations (usually layered in mid-summer), will root in a couple of months, but woody plants may take one or two years.

Two essential requirements for satisfactory results are the provision of good soil into which the stem can root, and the treatment of the shoots or stems in such a manner that the sap flow is reduced so as to encourage rooting.

If the stem or shoot is to be rooted directly into the soil surrounding the parent, it is advisable to work in plenty of peat and sand to encourage an open but moisture-retentive rooting medium. This should be kept moist at all times. As with other methods of layering, pots of a sandy peaty compost sunk into the ground at the correct place will ease transplanting of the young plantlet. Again, the compost must be kept moist.

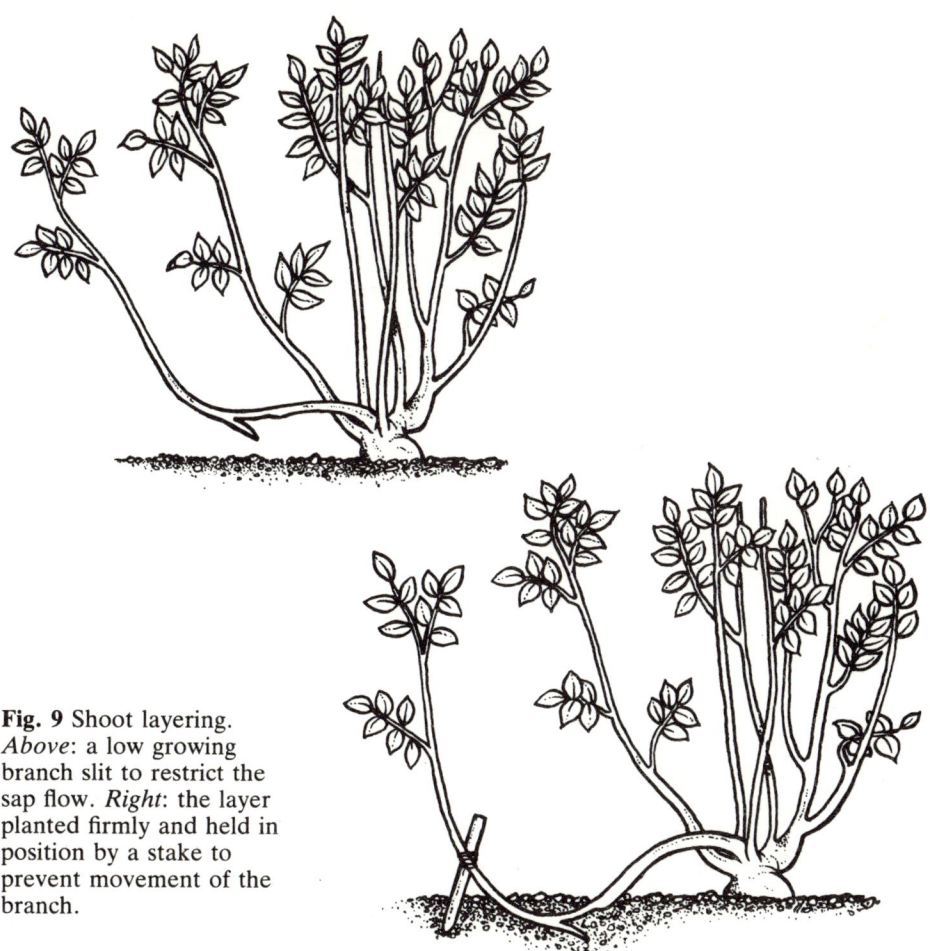

Fig. 9 Shoot layering. *Above*: a low growing branch slit to restrict the sap flow. *Right*: the layer planted firmly and held in position by a stake to prevent movement of the branch.

Select young one or two year old shoots, that are sufficiently flexible to be bent into position without breaking. Low-growing older shoots can sometimes be layered satisfactorily, but they will take longer to root and may produce poorly shaped specimens.

There are several methods of restricting the sap flow from the parent plant to the proposed youngsters. It can be done by notching or slitting about 2 cm (¾ in) of the underneath of the stem or shoot with a sharp knife at the point at which it is to root (Fig. 9); it can be given a twist by hand, which will break some of the plant tissues; or a circle of bark,

39

1 cm ($^3/_8$ in), can be carefully removed with a sharp knife. In all cases, it is advisable to carry out these operations at a node (stem or shoot joint) which is the point from which the new root and shoot growth is most likely to arise. Applying a proprietary hormone rooting powder to the treated areas also often proves beneficial in encouraging rooting.

To keep the area of stem or shoot to be layered securely in position it is advisable to hold it firmly in place with stones, or a piece of bent wire, forked wood, or a hooked tent peg pushed into the ground. The tip of the shoot or stem should also be tied to a vertical stake pushed firmly into the ground; this will help stability during layering as well as encouraging the plantlet to grow upright and form a well-shaped speciment plant. Always ensure that the layered stem or shoot is firmly in position and check from time to time to see it has not become loosened. Also, when it is anticipated that a layered stem or shoot will take some time to root, ensure that it is kept free of weeds by careful hand weeding.

Before transplanting a layered shoot or stem, ensure it is well rooted by carefully removing some of the soil round it with your fingers. If it has not rooted properly, leave it in position for another six to twelve months. If it is well rooted, sever the stem from the parent plant and leave it where it is for a week to ten days to get established on its own. After that, transplant it to its final position.

SERPENTINE LAYERING

Where a long shoot or stem of a climber, such as clematis, ivies or *Vitis coignetiae* (vine), is growing in a suitable position near the ground, serpentine layering may be employed to give more than one young plant from a single shoot or stem.

The same principles apply to serpentine layering as to that described previously, the only difference being that a number of 2 cm ($^3/_4$ in) long slits are made along the lower length of a young low-growing stem or shoot at the nodes, and each cut area is pegged down into the ground or pots. If this is carried out in late spring, the new plantlets should be ready for severing from each other and from their parent for transplanting in the early autumn. In cold areas of the country it is generally advisable to layer into pots, and then keep the young plants in a cold frame over winter and plant out the following spring.

AIR LAYERING

In some cases it is not possible to lower a stem or shoot far enough down for it to be rooted at soil level. Also, there are occasions when air layering is useful in reducing the height of an overgrown or straggly plant. It works most successfully on young fleshy growth, and is often used for the tender pot plants dracaena and ficus (rubber plant), but also for some forms of hardy magnolia, provided the tip end of a young branch is selected for propagation.

The technique of air layering is known to be thousands of years old and is sometimes referred to as Chinese layering. Originally it was carried out by using a split pot or box, both halves containing suitable compost, being put round the stem and tied into position and the branch supported to carry the additional weight. Nowadays, with the use of thin polythene sheeting, it is a much more simple operation.

Select a young shoot or stem, preferably not more than two years old and, with a sharp knife, either remove a ring of bark about 1 cm ($^3/_8$ in) in diameter, or make a V-shaped nick in the bark about 5 cm (2 in) long, at a distance of 15–30 cm (6–12 in) from the growing tip. Apply a proprietary hormone rooting compound to the cut area and, if you can get it, wrap very moist sphagnum moss around the treated area, to cover at least 7.5 cm (3 in) above and below it. Then wrap polythene film round the sphagnum moss and hold it in position securely with sticky tape or string around the stem at points above and below the moss to keep it air-tight. In this way no moisture is lost from the moss and after several months roots will appear.

If sphagnum moss is not available, the same general technique is used but dealt with slightly differently. After cutting or ringing the bark, secure one end of a 20–23 cm (8–9 in) piece of polythene sheeting to the 7.5–10 cm (3–4 in) point below the treated area in such a way that it forms a 'cup' or funnel shape. Into this 'cup' insert wetted proprietary soilless compost or moist John Innes seed sowing compost, so that the compost is tightly packed and covers the stem to a point at least 7.5 cm (3 in) above the treated part of the stem. Then secure the top piece of polythene securely round the stem or shoot and ensure that all the polythene sheet round the stem is air-tight.

The best time to carry out air layering is mid to late spring, and it may take months before roots are seen to be appearing through the rooting medium. During this period, care for the plants as you would normally, watering, feeding, controlling pests and disease, etc.

When the roots are plainly visible through the polythene, cut off the new plant just below the level of the rooting medium, remove the polythene (disturbing the roots as little as possible) and place the new plant in a pot of a suitable compost, such as a proprietary soilless one or John Innes potting compost no. 1. If the new plant is a hardy one from an outdoor parent, place it in a cold frame until it is well established and ready for planting out in its final position at the correct time of year. If it is a tender pot plant, the youngster should be kept slightly warmer than its parent, (by putting it into a propagating case, for example) until it is well established on its own roots. When this has been achieved, gradually harden it off to grow under normal conditions as required by its parent.

6 Stem Cuttings

Raising new plants from stem cuttings is one of the most popular methods of plant propagation. Essentially, it involves the removal of a length of stem from the chosen parent plant and placing it in a suitable growing medium and in the right conditions for it to produce roots, and so form a new plant. As with all methods of vegetative propagation, such as this, the new plant will be identical to its parent, whereas raising plants from seed can sometimes lead to variations (see Chapter 11). Stem cuttings are often the most convenient method for gardeners to give or exchange plants for propagation purposes.

Essentially there are three types of stem cuttings: hardwood, semi-hardwood (half-ripe) and softwood. Occasionally, when stem cutting material is in short supply, or *Vitis vinifera* (grape vines) are to be propagated, stem eye cuttings can be taken; also, with certain house plants, main' stem cuttings can be used.

Descriptions of the equipment to use for the propagation of cuttings, and how to use it to best advantage, are explained in Chapters 1 and 2.

HARDWOOD CUTTINGS
These cuttings are the easiest to take for raising new plants and many hardy trees and shrubs, including soft fruits, are propagated in this manner. Hardwood cuttings are those which are fully ripened shoot or stem growths of the current season and are usually taken in late autumn or early winter, though in mild winters they can be taken any time through

to early spring. Usually the parent plant is in a virtually dormant state, and the cuttings are not likely to suffer water loss due to transpiration (especially deciduous plants which will be leafless). For this reason, most hardwood cuttings can be rooted easily in the open ground, though evergreens (those plants which retain their leaves in winter) often root better if given the slight protection of a cold frame. In all cases, however, the production of roots is slow and it may be 12 months or more before the new young plants are ready to be transplanted to their final position.

In most instances the cuttings should be 15–30 cm (6–12 in) or more in length. The shoots to be used should be severed from the parent plant with a sharp knife and then be trimmed just below a node (a stem joint), so that the cut is completely clean and not jagged, which could adversely affect rooting and perhaps cause a source of disease (Fig. 10a). The prepared cuttings are then inserted in a 10–15 cm (4–6 in) deep V-shaped trench of well prepared, fertile soil in a sheltered part of the garden. A 2.5–5 cm (1–2 in) layer of coarse sand and damp peat at the bottom of the trench is helpful to rooting, and the cuttings should be set approximately half their length deep, and 7.5–15 cm (3–6 in) apart in rows 45 cm (18 in) apart. Replace the soil firmly by treading it down and, if it is loosened by frost during the winter months, re-firm it, as it is important to keep the cuttings steady and without movement which could destroy delicate new roots forming. Keep the beds of cuttings free from weeds. Water during the summer as necessary.

Although nodal cuttings, such as those described in the previous paragraph, are the type most often used for hardwood propagation, some people prefer to take heel cuttings, which are those with a strip of the bark of the old stem attached (Fig. 10b). Instead of cutting the stem or shoot from the plant, it is pulled off by hand to obtain the heel, but it is sometimes necessary to trim this smooth if the strip is jagged. Then proceed as for nodal cuttings.

Deciduous cuttings will probably have lost their leaves prior to being taken but, if not, it is best to remove them. With evergreen plants, remove carefully all the leaves from the length of stem that is to be inserted in the soil.

44

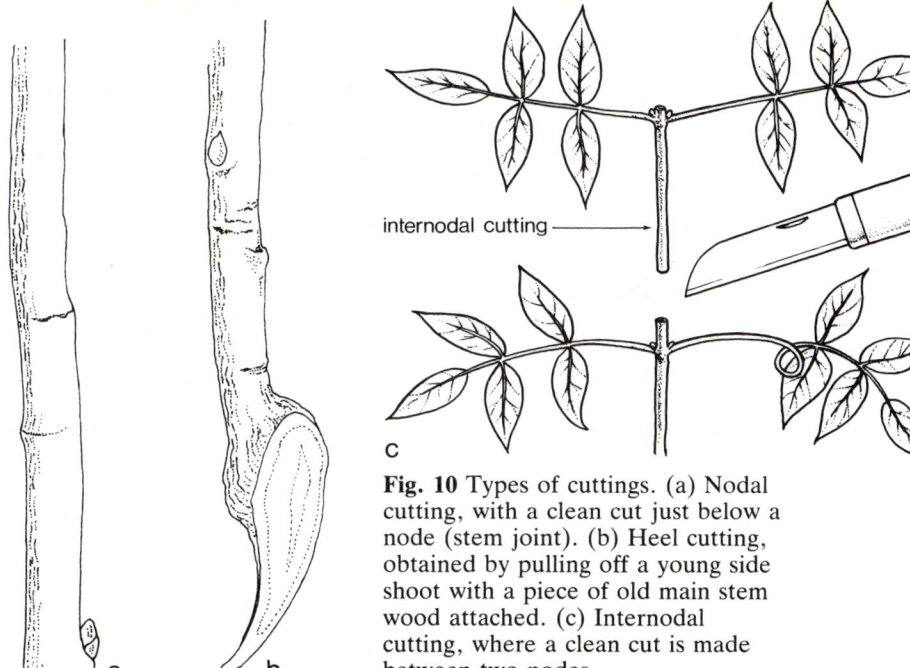

internodal cutting

c

Fig. 10 Types of cuttings. (a) Nodal cutting, with a clean cut just below a node (stem joint). (b) Heel cutting, obtained by pulling off a young side shoot with a piece of old main stem wood attached. (c) Internodal cutting, where a clean cut is made between two nodes.

a   b

Where the new plants are to be grown on 'legs' (a bare length of stem or trunk before side shoots arise), remove the buds, with your fingers, on the lower ends of the cuttings to prevent unwanted suckers and low growing shoots arising.

To encourage rooting of hardwood cuttings, particularly of any difficult subjects, it is useful to dip the prepared ends of each cutting into a hormone rooting powder or liquid of suitable strength immediately before inserting them into the ground. Follow the manufacturer's instructions as to how to use such a rooting aid.

Should it not be possible to plant hardwood cuttings in late autumn or early winter, due to inclement weather or soil conditions, it is quite possible to take the cuttings, tie them into bundles, and insert the bundles to half their depth in damp soil and leave them like this until the spring. They are then planted in rows as described previously.

SEMI-HARDWOOD (HALF-RIPE) CUTTINGS
Semi-hardwood cuttings are sometimes referred to as half-ripe ones because that is the stage of growth the stems or

shoots should be at when taken from the parent plant. The best way to describe them is as stems or shoots that have almost finished their growth period but which have not yet had time to ripen into hard, mature wood. Therefore, the most usual time to take semi-hardwood cuttings is from mid-summer to early autumn. Plants which can be propagated best in this manner are many deciduous and evergreen shrubs, particularly conifers, heaths and heathers (*Erica, Calluna, Daboecia* species and varieties) and clematis, which are often difficult to root at other times of year.

Because semi-hardwood cuttings will still have their leaves on when taken, be they deciduous or evergreen, they will tend to lose water through the leaves (by transpiration) and will wilt unless they are given the right conditions to overcome this problem. The best method is to insert the cuttings in a 50:50 coarse sand and damp peat mixture, or special rooting compost, in pots or boxes in a cold frame and keep the lights (glass covers) on tightly until rooting has taken place. Damping the cuttings daily with water from a fine rose on the watering can should be carried out and, during very sunny periods, it may be necessary to shade the glass with paper, hessian or wooden slatted blinds; but do not leave the cuttings in shade for too long or they will get drawn and lanky. Sometimes rooting can be speeded up by the use of soil-warming cables in the frame, or by putting the cuttings in a mist propagating unit.

Once the cuttings are seen to have rooted, by their sturdiness and appearance of new top growth, they are ready to be hardened off, by reducing heat and watering and the removal of the frame or propagating cover. After a week or 10 days, the plants should be planted into a soil compost, such as John Innes no. 1, either in pots or boxes.

Normally, the cuttings are left in their pots or boxes in a cold frame over the winter months and then planted out in their final positions later the following spring. As can be appreciated from the timing, semi-hardwood cuttings do not take so long to root and mature as do hardwood cuttings, but they do require a little more attention.

The cuttings are taken in a way similar to that described for hardwood ones, except that shorter lengths of shoots or

46

stems are used. With most shrubs prepare a 10–15 cm (4–6 in) long nodal cutting, with or without a heel, remove the bottom leaves and, with a dibber, insert the lower half of each cutting into the prepared compost and press it firmly into position with the finers. It is usual to pinch out the growing tip between the finger and thumb, as this tends to wilt. The use of hormone rooting powder or liquid is often beneficial and should be used according to the manufacturer's instructions.

In the case of plants which produce only small half-ripe shoots or stems suitable for cuttings, such as conifers and heathers, 5–10 cm (2–4 in) long tip cuttings of laterals are taken. These are treated in exactly the same manner, but the tips of the cuttings are not pinched out.

Occasionally, some plants, such as clematis, are thought to root better from what are called internodal cuttings (Fig. 10c). These are cuttings which are trimmed off between nodes instead of just below a node. They are then treated in a similar manner for rooting.

SOFTWOOD CUTTINGS
Softwood cuttings are usually those which are taken from the parent plant early in its season of growth so that the cuttings consist of immature succulent shoots or stems. These will root quickly—10 days to one month—under the right conditions, which involves giving them a close, warm, moist atmosphere and the correct rooting medium. Providing a suitable environment and maintaining softwood cuttings while they root requires rather more attention than that needed for hardwood or semi-hardwood cuttings, but it is one of the simplest methods of increasing stocks of greenhouse and houseplants, herbaceous perennials, certain half-hardy plants such as chrysanthemums, dahlias and pelargoniums, and a variety of sub-shrubs (low-growing shrubs with soft stems).

Softwood cuttings of hardy outdoor plants are best taken early to mid-summer, those of half-hardy plants in late autumn or winter, and those of tender greenhouse or houseplants at any time of year. If the plants you wish to increase do not possess any suitable young growths, these can often be induced by cutting back old stems or shoots which will encourage the production of young and soft side shoots. Cuttings referred to as basal ones are young shoots taken direct from the roots and not from old stems.

Whenever possible, softwood cuttings should be firm but not hard and 5–10 cm (2–4 in) long, non-flowering side shoots; they are usually of the nodal type, that is, trimmed with a sharp knife or razor blade to just below a node. If the parent plants are very small, such as alpines, then it may be necessary to take cuttings as small as 2.5 cm (1 in) in length. Sometimes internodal cuttings are preferred for certain plants, such as hydrangeas, fuchsias and lavenders. Where a plant may be hollow-stemmed, such as is the case with some herbaceous plants, heel cuttings make for quicker and easier rooting; these are taken as described previously for hardwood and semi-hardwood cuttings. *Dianthus* species (carnations and pinks), are often increased by what are called pipings; these are non-flowering side shoots growing from the main stem which are pulled off by hand so that they slide out from just above a pair of leaves (Fig. 11).

As all softwood cuttings tend to be sappy and will have their leaves on, they will wilt rapidly. If taking them in quantity, place them in a polythene bag or a lidded container to prevent evaporation of moisture. Also have the containers and rooting medium prepared in advance. Some-

Fig. 11 Pipings of carnations and pinks. (a) Non-flowering side shoot pulled out from the main stem by hand. (b) The pipings set around the edge of a pot containing a sandy peaty compost.

48

times such cuttings can 'bleed'—exude sap from the cut stem—but this can usually be arrested by dipping the ends in charcoal powder. Equally, the use of a hormone rooting liquid or powder often speeds up rooting and is helpful with difficult plants. If it is of the type which contains a fungicide, so much the better; because of the humid, warm type of environment used for rooting softwood cuttings, it is ideal also for the spread of diseases. (It is equally important always to remove any dead or decaying leaves or cuttings noticed during the rooting period.)

The best rooting medium for softwood cuttings is one consisting of equal parts of coarse sand and damp peat; alternatively, a proprietary cuttings compost, such as John Innes, may be used, or a seed sowing compost. In the two latter cases, a sprinkling of silver sand over the surface, so that some of it falls into each of the holes made to take the cuttings, will help rooting.

The container used for the rooting medium will depend on how many cuttings are being rooted and what equipment is available. It will also, to a certain extent, depend on what plants are being propagated and the temperatures the cuttings require as a result. Greenhouse and house plants, for example, usually need temperatures of 13–18°C (55–64°F), though sometimes they may be lower or higher depending on the genus and species. Hardy plant cuttings will also benefit from more heat than their normal growing temperature, but it is important that this should be bottom heat only, that is, in the area of the rooting medium but not around the cuttings themselves where it would cause the leaves and stems to dry out, or encourage shoot growth before roots have been formed to support it. Furthermore, adequate humidity is important to prevent the cuttings from wilting, and shade from hot sun is required for the same reason, for at least the first few days.

If large numbers of softwood cuttings are to be rooted, a wooden box, or a propagator or cold frame, will be required, with or without bottom heat according to the time of year and temperature requirements. For a few cuttings only, they can be placed individually in pots, or several can be inserted per pot, preferably round the edge, where they seem to root more readily, especially if the pots are of clay or earthenware. To prevent growth checks when planting out, peat or peat/soil pots can be used.

As it is essential with most plants propagated in this manner that a humid atmosphere be maintained, either cover boxes and pots with polythene sheeting or bags secured firmly in position, but with the polythene kept clear of the cuttings by stakes or wires, or keep the covers on propagating cases and the lights firmly closed on frames. Plunging pots or boxes in propagating cases or frames half filled with damp peat on a layer of sand is another way of ensuring adequate humidity. Syringing the cuttings with water may occasionally be necessary. If you are fortunate enough to have a mist propagating unit, this will greatly assist in rapid rooting and prevent transpiration losses. Exceptions to this rule are soft hairy leaved plants and some alpines, which can rot quickly in very moist conditions, and certain shrubs, such as cistus and greenhouse azaleas, also chrysanthemums and dahlias, which seem to prefer a more buoyant atmosphere. These are best raised on warmed greenhouse benches without being covered.

As mentioned, it is equally important to shade softwood cuttings from hot sun or bright light during the first few days to prevent wilting. This can be done with paper, or wooden slatted or other types of blinds. In general, the cuttings should be kept out of direct sunlight until they are rooted and growing away strongly, but they should not be deprived of ordinary light unnecessarily, as the leaves require light to manufacture essential plant foods.

When all the equipment is ready and to hand, prepare the cuttings for insertion into the compost. In general, most softwood cuttings should be inserted to about one third of their length, therefore remove the leaves from this lowest portion of each stem or shoot. Ensure the base of the cutting has been cleanly trimmed with a sharp knife or razor blade and, if a flowering shoot has had to be used, pinch out the tip with the finger and thumb. Then, with a dibber or pencil, make a hole for the cutting in the compost, place it carefully in position and firm the compost round the stem with your fingers, dibber or pencil. Each cutting should be 5–8 cm (2–

3 in) apart from its neighbours. When all the cuttings have been inserted, water the compost to help it settle and syringe the leaves lightly with water (except woolly leaved plants as mentioned earlier). If in containers, place these in the greenhouse, frame, home, or wherever they are to be kept until the cuttings have rooted, and treat them as described previously, being careful to keep them in a close humid atmosphere and protected from strong sunlight.

Looking after softwood cuttings until they have rooted is important if good results are to be obtained. They should be examined daily and any that are obviously diseased should be removed, also any dead leaves, to prevent the spread of fungal infections. If necessary, water the rooting compost, also the bed of soil and peat in which containers may have been plunged, and syringe lightly with water if the cuttings are wilting and seem at all dry. Also wipe the glass or polythene covers clear of moisture droplets so that water does not fall on the cuttings. Keep the shading in position to prevent hot sun striking the cuttings for the first few days until they look healthy and firm, then gradually withdraw the shading until they are fully exposed.

As it becomes apparent that the cuttings are rooting, harden them off gradually by reducing temperatures, giving ventilation for short periods and by removing polythene or glass covers or frame lights. Finally, the covers can be removed completely and the new plantlets hardened off to their normal growing temperatures before they are potted on (if greenhouse or houseplants) or planted out (if they are hardy plants). During this interim period, giving dilute feeds of a liquid fertilizer can be beneficial.

Cuttings which have not been taken until the autumn, for example pelargoniums (geraniums) and violas, will require protection throughout the winter in cool but frost-free positions, such as in a garden frame, in which case the frames should be kept closed during the cold months and ventilation gradually increased in spring before they are fully hardened and ready to be planted out. Water sparingly during the winter months and, when frost is expected, cover the frames at night with hessian, sacks, old carpets or such like to maintain some warmth.

EYE (BUD) CUTTINGS

When a plant is not producing enough suitable shoots for cuttings to be taken as described previously, it is possible to propagate them by eye (bud) cuttings. This is a slow method and usually requires a minimum temperature of 15–16°C (59–61°F), so a propagating case is advisable to provide the warm, humid conditions required.

Take a hardwood leafless stem and cut it into sections, so that each section is cut just above a bud, preferably with a sloping cut. Trim the basal portions to a length of about 2.5 cm (1 in) below the bud and insert these into a 50:50 peat and sand compost in individual pots. Plunge each pot into damp sand and peat in the propagating case. Treat them as you would softwood cuttings.

A variation of this method is used to propagate *Vitis vinifera* (grape vine). The method is similar to that described in the previous paragraph but, instead of inserting the eye cuttings vertically, remove a strip of wood from the cutting on the opposite side from the bud and lay the bud piece horizontally on the rooting medium, so that the bud is just above compost level. Secure the cutting each side of the eye with bent wires stuck into the compost.

Rooting will probably have taken place with both these types of eye cuttings when healthy top growth appears from the bud. This is the time to harden them off slightly and re-pot each new plant into John Innes potting compost no. 1. Repot a second time into a size larger pot containing no. 2 compost, when the first is filled with roots, and then harden the plants to their normal growing temperatures and pot on to larger pots or plant outdoors as required.

MAIN STEM CUTTINGS

These are a means of increasing certain houseplants, such as cordyline, dieffenbachia, dracaena and philodendron. The top 7.5–10 cm (3–4 in) is cut off with one mature leaf and used as a single cutting. The leaves on the remainder of the stem are then removed and the stem cut into sections, each with two or more buds. These short lengths are inserted vertically in the rooting medium and are treated as are softwood cuttings.

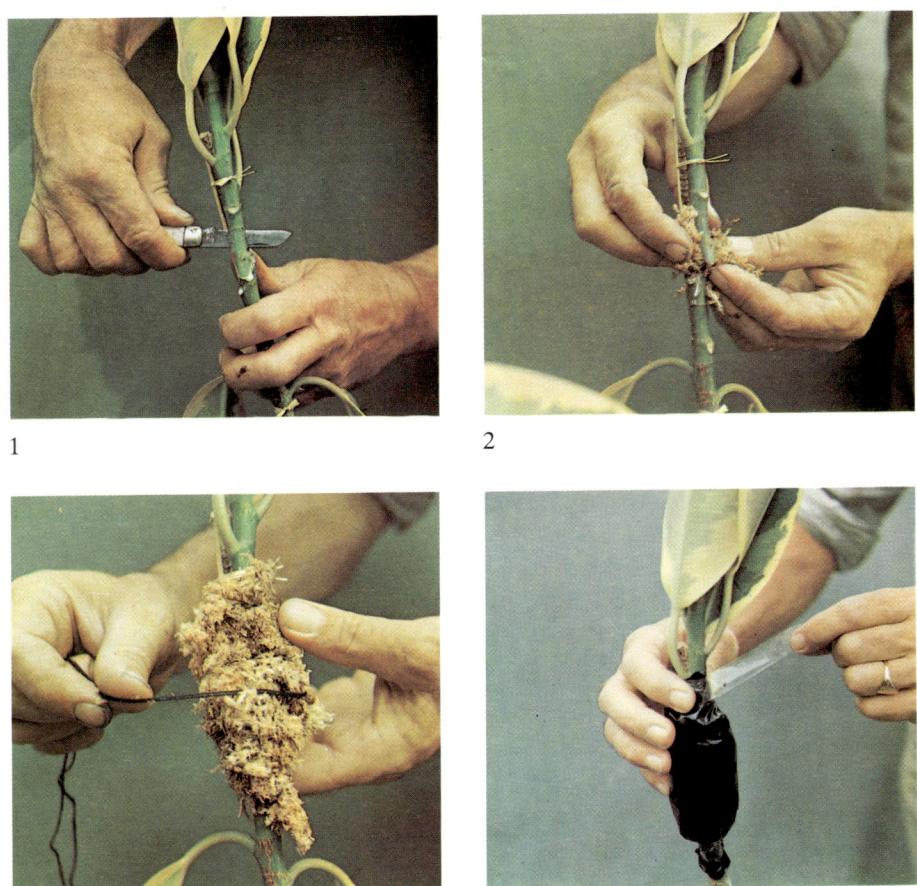

1

2

3

4

Air layering *Ficus elastica*, showing the making of a main stem cut, inserting sphagnum moss into the cut and well around it and the stem, before wrapping the treated area with polythene and sealing the ends with sticky tape.

When raising cuttings in pots
covered with polythene bags, it
is important to keep the
polythene bag away from the
leaves, by supporting the bag
with a wire hoop, as illustrated
here, or with twigs.

7 Root Cuttings

Any plants which produce suckers naturally, such as *Rubus idaeus* (raspberries), and their near relatives *R. fruticosus* (blackberries), and *R. loganobaccus* (loganberries), can be increased easily by root cuttings. So also can a number of hardy herbaceous perennial plants which have fairly fleshy roots, such as verbascum, eryngium, papaver, romneya and anchusa, as well as vegetables such as *Crambe maritima* (seakale), and *Cochlearia armoracia* (horseradish). In some instances, plants with thinner fibrous roots can be propagated from root cuttings, and these include phlox and gaillardia.

Plants which have been grafted or budded on to rootstocks should only have their roots increased by root cuttings if a supply of rootstocks is required for increasing plants by grafting or budding (see Chapters 9 and 10). This applies usually to plants such as roses, syringa, and ornamental and edible tree fruits.

Root cuttings are best taken during the plant's dormant period, which means any time during the winter months when the weather is reasonably mild and the soil conditions suitable.

If only a few cuttings are required, carefully scrape away soil from around the roots of the chosen parent plant, with a hand fork and the fingers. If a number of cuttings are required, it is easier to lift the plant from the ground completely, but do this carefully with a border fork to create the minimum of root disturbance.

When taking fleshy root cuttings, select those roots which

are about 1 cm ($^3/_8$ in) in diameter and cut them into 7.5–10 cm (3–4 in) lengths. In order not to confuse which way up the root cutting should be placed in the rooting medium, make a horizontal cut at the top end (nearest the plant) and a sloping cut at the bottom end. These prepared cuttings are then inserted into pots or boxes containing a compost of equal parts of sterilized soil, coarse sand and moist peat, or a proprietary cuttings compost such as John Innes (Fig. 12a). Make holes of the required depth with a dibber and insert the sloping cut end into the hole, so that when the cutting is in position its top end is just covered with the rooting medium, or a 1 cm ($^3/_8$ in) layer of coarse sand if you prefer. The pots or boxes are then stood in a cold frame or under cloches for the remainder of the winter, watered occasionally if the rooting medium dries out at all and new plants should form in the spring; this will be noticed when shoots and leaves appear above ground. By early summer, these youngsters should be ready to be planted in rows in a nursery bed of well-prepared soil until such time as they are required for planting in their permanent positions in the autumn.

It is quite possible, also, to get fleshy root cuttings to grow in the open ground in gardens with a mild climate or where there is space for a nursery-bed in a sheltered corner. During inclement weather they can be protected with cloches. Treat the cuttings exactly as described above, but space them 15–20 cm (6–8 in) apart so that transplanting in the spring is not necessary. Also ensure the ground is well-prepared, forking in coarse sand and damp peat to improve it if necessary. Mark the rows of cuttings, or place a label by each individual cutting, so there is no danger of damage to them, or putting other plants over them by mistake while they are still below ground. At all times keep the nursery areas weed-free.

Fibrous rooted cuttings are treated very similarly but, instead of inserting them into the rooting medium vertically, they are laid horizontally on it (Fig. 12b). For this reason it is not necessary to differentiate between the top and bottom of the cutting by making a sloping cut. It is easier to use boxes rather than pots for these types of cuttings.

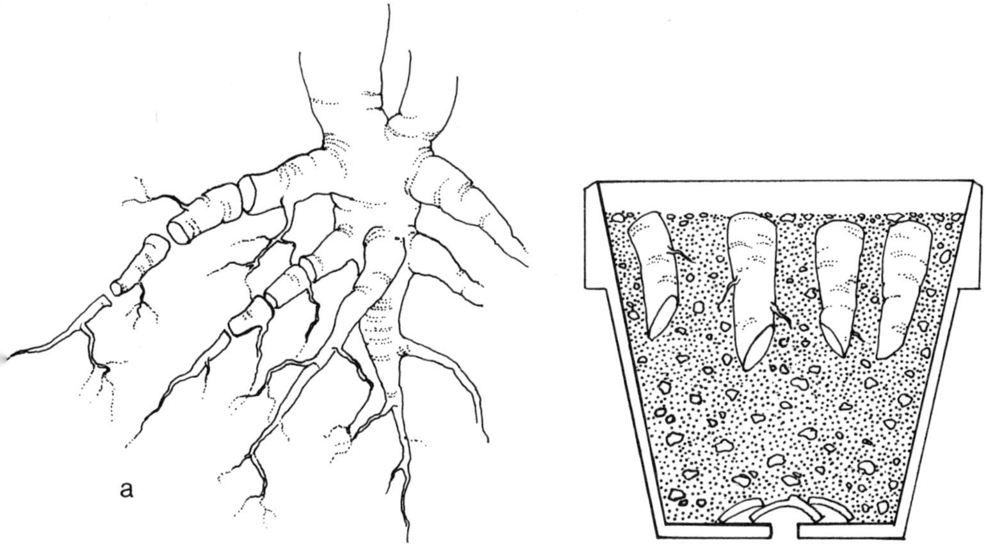

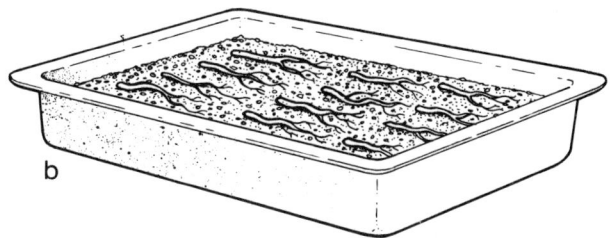

Fig. 12 Root cuttings. (a) Cut
fleshy roots into 7.5–10 cm
(3–4 in) lengths, the lower
end with a sloping cut; insert
in a pot of compost, sloping
end down. (b) Fibrous root
cuttings are laid horizontally
on compost in boxes and
covered with a layer of the
same compost.

The cuttings are covered with 1 cm ($\frac{1}{8}$ in) of sand or compost and the boxes stood in unheated frames or under cloches. Fibrous root cuttings can be rooted in the open ground
by laying them in a shallow trench and covering them with
about 5 cm (2 in) of good soil, or a mixture of equal parts
of peat, sand and soil. If the ground is heavy, it is advisable
to put a layer of similar rooting compost beneath the cuttings
as well.

In all cases, the parent plants should be replanted as soon
as the root cuttings have been taken; likewise, the cuttings
should be inserted in their rooting medium as quickly as
possible.

8 Leaf and Leaf-bud Cuttings

A few greenhouse and outdoor plants can be increased by leaf and leaf-bud cuttings. The former involves the use of a leaf, sometimes with a leaf stalk and bud, the latter requires also a small portion of the stem.

LEAF CUTTINGS
Raising new plants by leaf cuttings, with or without a bud attached, is the method used most frequently to propagate certain greenhouse and houseplants. Among the most common of these are saintpaulia, peperomia (pepper plant), gloxinia, *Begonia rex* and other begonias grown for their foliage display, sansevieria, streptocarpus, *Tolmiea menziesii* (pig-a-back plant) and *Asplenium bulbiferum* (spleenwort fern).

In the case of tolmiea and asplenium, small plantlets are frequently produced quite naturally on the leaves. These can either be pegged down alongside the parent plant in the same pot (if space allows), be pegged into a small pot of rooting compost alongside, or, if the plantlets are growing strongly, they can be cut off with a sharp knife at the base and be inserted directly into pots of rooting compost. Whichever method is used, the plants need to be kept in a temperature of around 15°C (59°F) and be watered as necessary. Once the pegged-down plants are growing away strongly on their own roots, they are severed from their parents. In all cases the new plants are then potted up at intervals into larger pots containing stronger fertilizer grades of compost until they are in the size of pot required for the

58

mature plant, usually of the 13–18 cm (5–7 in) diameter size, with a compost equivalent to John Innes no. 2.

When propagating the smaller leaved plants, such as gloxinia, saintpaulia (Fig. 13a) and peperomia, select healthy adult leaves and cut them from the parent plant at the base of the leaf stalk. Insert the leaf stalks into pots or boxes containing a 50:50 sand and moist peat compost (by making holes with a dibber or pencil), so that the full length of the stalk is buried and the base of the leaf just rests on the rooting medium. Roots and shoots will then grow from the base of the leaf stalk. It is also possible to root these plants from leaf cuttings inserted in a small container of plain water, but care must be taken to ensure that the actual leaves are kept out of the water. Once the leaf stalks are

Fig. 13 Leaf cuttings. (a) Sever a leaf stalk from the parent plant of an African violet and insert it in a pot of peaty sandy compost; a new plant will appear from the base. (b) Cut the main veins of a *Begonia rex* leaf, lay it on specially prepared compost, pin or weigh it down flat, and new plants will arise from each cut area.

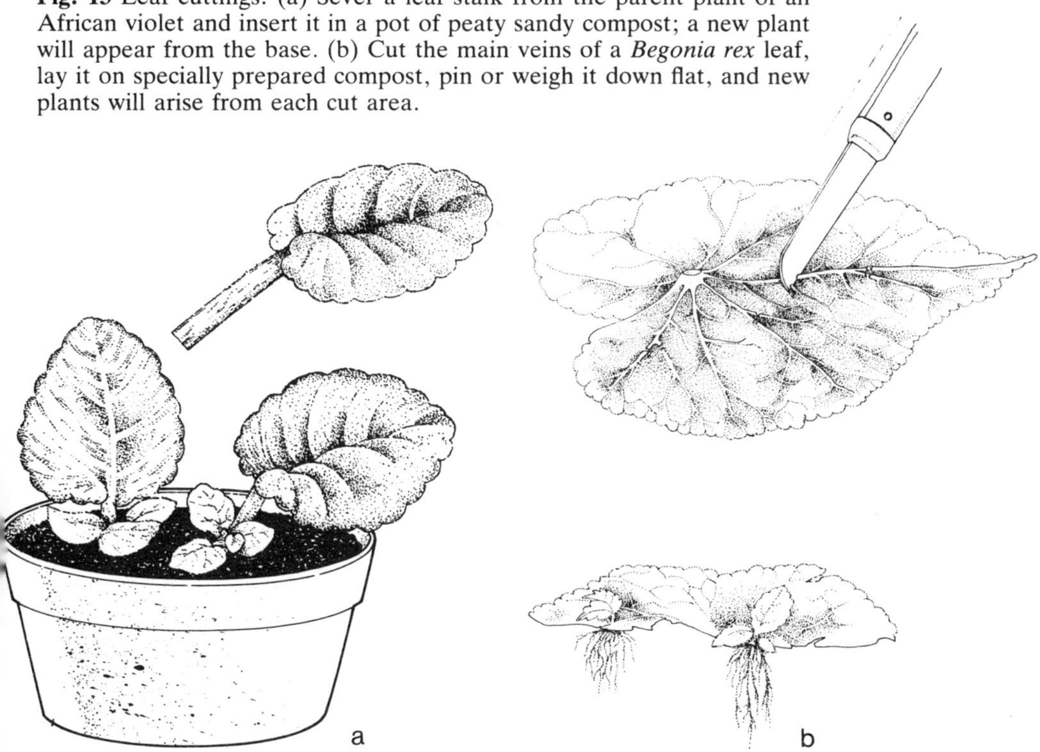

a

b

seen to be rooting freely, each new plantlet is potted up carefully into a small pot of potting compost, great care being taken not to damage the fragile roots.

Larger leaved plants, such as begonias, streptocarpus and sansevieria, can have their leaves cut into sections to produce more than one new plantlet from each leaf.

There are two methods of increasing *Begonia rex*. Either cut the main veins on the undersurface of the leaf (Fig. 13b), lay it flat on the rooting medium (50:50 coarse sand and damp peat in pots or boxes) and hold each cut area in position with pebbles or U-shaped pieces of wire. New plants will then grow from each cut section. Alternatively, cut the leaf with sharp scissors into small main vein sections and insert the vein ends vertically into the rooting medium. New plantlets will form from the base of each leaf portion.

Streptocarpus and sansevieria leaves are best cut into separate sections, with the basal piece of each section trimmed into a wide angled V-shape so that the main vein can be inserted in the rooting medium. Each piece of leaf treated in this way should be 2.5–5 cm (1–2 in) in length. As with the other plants described previously, new plantlets will form from the base of each section.

In all cases where leaf cuttings are taken, rooting will be quick and free if the cuttings are kept in a warm, moist atmosphere, such as that provided by a propagating case. Bottom heat is particularly valuable and, as these plants are all greenhouse or houseplants, an air temperature of about 15°C (59°F) is also necessary. If a propagating case is not available, and there is no greenhouse in which a temporary propagating unit can be created, covering the pots or boxes with polythene and standing them near a radiator in the house will often produce satisfactory results.

When the plantlets have produced roots and shoots harden them off somewhat by removing the top of the propagating case or polythene bags for a few days, then pot each into a small pot containing a mild potting compost. Handle the plantlets with care so as not to damage the young growth, making the holes in the compost with a dibber or the fingers. Gently firm each plantlet into position and water the compost to settle it around the roots.

60

As described for increasing tolmiea and *Asplenium bulbiferum*, pot up the plants as they grow into larger sized pots and stronger compost until they are in the final pot size you require.

LEAF-BUD CUTTINGS

These are not dissimilar to eye cuttings described in Chapter 6. The plant most frequently propagated by leaf-bud cuttings is *Camellia japonica*. In this instance, the leaf and leaf stalk are cut from the parent plant with a small piece of stem attached which bears a growth bud. The best time to take such cuttings is while the plants are growing strongly, generally in mid-summer.

The stem portion is trimmed with a razor blade, and then the bud and leaf stalk are inserted in the growing medium just as you would a leaf cutting. To encourage the production of roots and shoots from the buried bud, the same close environmental conditions are required, such as those provided by a propagating case.

Once rooting has occurred and a new shoot appears, harden off the plants, put each into a separate pot of potting compost and grow on in gradually cooler conditions, until they are ready to be over-wintered in a cold frame. In spring, plant out in rows in a nursery-bed, keeping it weed-free and watered as necessary, and the new plants will be ready for transplanting to their final positions the following autumn.

9 Grafting

Grafting, and budding (see Chapter 10), are more complict-
ed methods of vegetative propagation than those described
in previous chapters. They are not widely used in most
gardens, though they are a common method for nurserymen
to increase stocks of plants in quantity. In most instances,
they are used to produce new plants which are difficult to
raise from division, layering or cuttings, or which do not
come true to type from seeds. The type of plants most
frequently grafted are edible and ornamental top fruits, such
as apples, crab apples, pears, plums, peaches, quinces, nectarines and ornamental cherries, and some shrubs, such as
syringa (lilac), hybrid roses and laburnum.

The term grafting means to attach a section of the plant
variety required—known as the scion—to another closely
related plant, usually the common species of the genus,
which is growing on its own roots, and which is called the
rootstock. The scion and rootstock grafted together then
unite to form the desired plant, with the scion producing
the quality and type of flowers and fruit required and the
rootstock giving the new plant the strength and vigour it
needs to grow freely and perform well. Occasionally, scions
of the chosen plant are grafted on to rootstocks of other
closely related genera within the same family; for example,
syringa may be grafted on to ligustrum (privet), sorbus on
to crataegus (ornamental thorn), and cytisus (broom) on to
laburnum. In the latter case, if *Laburnum anagyroides* and
Cytisus purpureus are grafted together, a graft hybrid known
as × *Laburnocytisus adamii* is produced, which bears pink,
yellow and purple flowers all on the same tree.

What is absolutely essential in grafting, is that the scion and the rootstock are compatible and that what are known as the cambial layer of cells of each (which are just below the bark) unit together to form a firm and strong joint so that a healthy and long-living plant is the result of the union.

In some cases considerable research has taken place to ascertain which rootstocks are most suitable for grafting purposes. This applies particularly to those for top fruits and roses. In the former cases, the vigour of the rootstock can have a considerable bearing on the ultimate size and fruiting performance of the grafted tree. Therefore, stocks have been carefully selected to give uniform results. Perhaps the most popular and well-known apple stocks are the Malling and Malling-Merton range, which can be selected for the variety to be grafted and the size and vigour required for the ultimate tree. For example, 'M7', 'M9', 'M26' or 'MM106' are most suitable for smaller growing specimens, such as cordons, bushes and dwarf pyramids, whereas stocks designated 'M16' or 'M25' are vigorous enough to produce large standard type trees with early cropping capacity.

Plums are usually grafted on to sucker or seedling rootstocks or 'St. Julien A', and cherries on to the wild *Prunus avium* or on a rootstock known as 'F12.1'. Certain pear varieties can sometimes be difficult to graft on to the quince (cydonia) rootstocks used, such as 'Malling Quince A' and 'Malling Quince C'. In this case, a double graft is required, using a variety that is compatible to both the rootstock and the desired scion to act as a short 'bridging gap' between the two. Both grafts can be carried out at the same time.

For grafting roses, the most common stocks are the wild dog rose (*Rosa canina*), which can be found in hedgerows, *R. laxa* and *R. multiflora*. However, if prunings of rambler roses are used as cuttings to form roots, these can quite readily be used as rootstocks for all types of roses.

For rhododendrons, *Rhododendron ponticum* makes a good rootstock for selected varieties which are difficult to increase except by grafting.

Often one of the problems with grafted plants is that the rootstock will throw up sucker growths. Unless these are removed completely, by pulling them away from the part of the roots from which they have arisen, they may swamp the plant and will certainly reduce its vigour. On the other hand, however, they are a useful source of fresh rootstocks for further grafting or budding.

The rootstocks should be purchased from a nurseryman or be prepared in the autumn. Plant them in a well-cultivated nursery-bed, setting them 30 cm (12 in) apart in rows about 60 cm (2 ft) apart. It is also advisable to start preparing the scions during the winter months. Cut about 15 cm (6 in) long shoots from the plant chosen to be propagated, tie them in bunches, and put these the right way up in shallow soil trenches in a cool shady place; they will then remain dormant but plump longer than if left on the tree.

The actual act of carrying out the grafting process is best done in early to mid-spring. There are a number of different methods of grafting but it is proposed to describe only those which are of most use to gardeners. In all cases, however, the grafted unions should be tied together with raffia or plastic tape and be sealed with a proprietary grafting wax until the scion starts to produce new shoot growth and leaves. This is usually within a few weeks. Once you feel certain the union has taken place satisfactorily, very carefully remove the tying material so that it does not cut into the swelling main shoot. If you are in any doubt as to whether the union has formed a complete callus, it is wise to re-tie it and leave it for a further few weeks.

SPLICE (OR WHIP) GRAFTING
This is the simplest form of grafting and is best used where the scion wood and rootstocks are thin, such as with cytisus (broom) and roses. With a sharp knife or razor blade, make an oblique cut at the top of the rootstock. Then make a similar oblique cut at the base of the scion wood (Fig. 14a). Hold the two cuts firmly together and bind with the tying material to hold them in position. Cover with tape or grafting wax.

WHIP AND TONGUE GRAFTING
This is the most commonly used of all forms of grafting for those trees and shrubs that are best propagated by grafting.

64

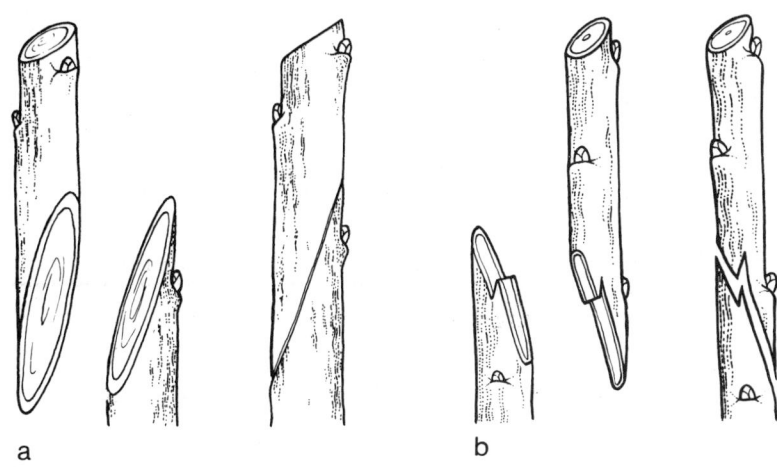

a   b

Fig. 14 Grafting. (a) Splice (or whip) graft: using stems of the same thickness, make oblique cuts in the scion and stock, hold the two closely together and bind into position. (b) Whip and tongue graft: first make oblique cuts in the stock and scion, then make a downward cut in the stock and an upward one in the scion and insert the latter over the former and tie firmly together.

It is very similar to splice grafting, but more reliable. The height at which this type of graft is made is usually 23–25 cm (9–10 in) above ground level. First cut off the tip of the rootstock, and remove any side shoots, buds or leaves. Then, with a sharp knife, make an upward sloping cut which is about 5 cm (2 in) long. Next take the scion wood, and make a similar oblique cut of the same length (Fig. 14b). If possible, ensure that the diameter of the stock and scion woods are similar, as this will ensure that both the cambium layers will be placed against each other. If either is larger than the other, ensure that the cambium layer and bark on one side at least match together. Then make an upward cut, about 1 cm ($^3/_8$ in) long in the centre of the scion wood and a similar downward cut in the rootstock. These two 'tongues' are fitted one on top of the other so that the graft is held firmly in the correct position. Bind and seal the union in the previously described manner, making sure that the whole of the cut areas are covered. The interlocking of the 'tongues' makes it much easier to handle the tying operation than with a simple splice graft.

SADDLE GRAFTING

Saddle grafts are most used to propagate rhododendrons and certain other shrubs which are difficult to root from cuttings. It involves removing the tip of the rootstock and then cutting it upwards with oblique strokes on either side so that it forms an inverted V-shape. A similar V-shaped wedge of wood is cut from the base of the scion, so that it sits comfortably, like a saddle, over the top of the rootstock cut. Tie the two pieces in position, wax the tied area, and treat as for other methods of grafting described previously.

If this type of grafting is carried out in the warmth and humidity of a propagating frame, with the rootstocks growing in pots, better and quicker results are often obtained than when done in the open. It will be necessary to harden off the new plants before setting them outdoors.

CROWN (RIND) GRAFTING

Where it is wished to rejuvenate old fruit trees, or to obtain a 'family' apple or pear tree (which consists of grafting several varieties of apples or pears together on each tree), the simplest form of grafting is the crown or rind method.

Cut back all but one or two of the main branches of the existing tree to within about 60 cm (2 ft) of the trunk and remove all side shoots. (The one or two branches left will help keep the stock growing freely and can either be removed completely or be grafted similarly a year or two later.) With a sharp knife, make a downward 5 cm (2 in) vertical cut from the top of the beheaded branch, cutting through the bark and into the wood below. Carefully lift back the flaps of the bark from the wood and insert the scion wood, which has had a 5 cm (2 in) long oblique cut made at its base. The flaps of bark are then put back over the scion. Repeat this treatment if the branch being grafted is a largish one, so that each scion is about 7.5 cm (3 in) apart. See that the scions are securely in position, with their cut surface pushed firmly against the exposed wood of the branch, and that the bark flaps are in position over them, then bind them securely with raffia or tape and cover not only the grafted areas but also the cut-off end of the branch with grafting wax, so that a complete seal is achieved. Fi-

66

nally, cut off the tips of the scions, just above an outward pointing bud, with the cut sloping away from the bud.

Crown or rind grafting is also a useful method to use to propagate other trees and shrubs where there is a fairly large difference in diameter between the rootstock and the scion, and splice (or whip) or whip and tongue grafting is unlikely to be successful because the cambium layers cannot be brought into good contact with each other.

BRIDGE GRAFTING

Although this is a method of grafting, it is used for repairing trees and woody stemmed shrubs and not for propagation purposes. When the bark and cambium layer has been removed or badly damaged all round the circumference of the trunk, the tree or shrub is likely to die as the plant's food supplies will not be able to flow between the roots and shoots. The same applies if a large branch is damaged in a similar fashion.

To remedy the problem, scions taken from the same tree or shrub, or from the same variety, have 5 cm (2 in) long slanting cuts made at the tops and bottoms and these are inserted into 5 cm (2 in) long vertical cuts made in the healthy bark either side of the damaged area—just as is done with crown (or rind) grafting. Make sure the pieces of scion wood are long enough to go over the damaged parts and well up into the healthy bark. Place the bark flaps over the scions, tie and seal. As previously, the pieces of scion should be spaced at about 7.5 cm (3 in) intervals. These 'bridges' will ensure a free flow of sap and will grow and strengthen the tree or shrub to give it a healthy life-span.

10 Budding

Budding is, in fact, a method of grafting but the technique is usually treated as a separate method of vegetative propagation because only a single bud with a small piece of bark from the scion (the plant to be increased) is used, whereas in grafting a length of stem with several buds is involved. It is employed by gardeners particularly for increasing roses, which do not always root readily from cuttings or which are weak if grown on their own roots. It can also be used for propagating ornamental and edible top fruits such as apples, pears, plums, cherries, peaches and nectarines.

The principles of budding and grafting are identical: namely, that a piece of the chosen variety of tree or shrub to be increased is attached to a suitable rootstock (the plant which acts as the 'host' to provide a suitably vigorous root system and encourage the formation of the size and type of new plant required). As with grafting, it is essential that the cambium layers of the scion and rootstock woods are kept firmly in close contact to establish a strong and healthy union. The bud is not involved in the actual union of the two plants, but is there to produce the shoots which will form the framework of the new plant.

Another slight difference between budding and grafting is that budding should take place in mid-summer when plant sap is flowing freely, whereas in the latter case it is carried out in early to mid-spring. The same disadvantages occur with budded plants as with grafted ones, particularly the throwing up of root suckers. These must be removed, preferably by pulling them off from their source by hand.

PREPARING THE STOCKS

The required rootstocks (see Chapter 9 for stocks most commonly used) should be purchased or obtained from rooted cuttings or seedlings the autumn prior to budding. They should be planted in rows in a well-prepared nursery-bed; set the plants about 30 cm (12 in) apart in rows spaced at 60–90 cm (2–3 ft) intervals.

In general, the stocks used for budding should have main stems of smaller diameter than those used for grafting, or the buds should be inserted on thinnish stems or shoots of the rootstock. It is rare for a satisfactory bud and stock union to take place when the chosen part of the stock is old and woody.

After planting, prune the stocks to the required height. In the case of most roses this is to about 15 cm (6 in), or 90–150 cm (3–5 ft) for half or full standards, and for most short growing and trained top fruits 30–60 cm (1–2 ft). Budding is not recommended as a suitable method of propagation for large standard fruit trees, as the unions forming the head framework often do not have the strength to withstand gale force winds.

The following mid-summer, before commencing budding, remove all the side growths from the stems of half and full standards and leave only three or four shoots at the tops into which the buds can be inserted. It is also wise to check whether the bark can be lifted freely. Do this by making an L-shaped cut with a sharp knife and see whether the bark can be lifted back cleanly from the wood. If so, all is well; if not, wait a week or two and try again. For easy, successful budding it is essential the bark lifts without difficulty.

PREPARING THE BUDS

The buds should come from lengths of scion wood selected from ripened current year's growth. The top buds will probably be too small and immature to use and should be cut off, while the lower ones should also be discarded as they are likely to be over-ripe. This will ensure there are a number of buds from the centre part of the scion stem which will be in suitable condition for budding. With a sharp knife remove the leaves growing just below each bud, taking care to leave part of the petioles (leaf stalks), which are used to help in the handling of the buds.

Carefully label each piece, or bundles of pieces if they are of one variety, insert them the right way up in a jar containing a little water and leave them overnight.

THE TECHNIQUE OF BUDDING

Once both the stock and bud wood (scions) are prepared, you are ready to carry out the technique of budding. The type most often employed, and recommended as being the most suitable for small scale propagation in the garden, is known as shield or T-budding. The operation is best carried out on a cloudy, preferably moist, sort of day to prevent stocks and scions drying.

T-shaped cuts are made at the correct levels on the rootstocks. For bush roses this is done just below ground level, by scraping away with a trowel some of the soil so that budding can be carried out as near the stock roots as possible. With half and full standards the cuts are made on the lower parts of the stems or shoots left after pruning the previous autumn. For top fruits, the cuts are usually made 15–30 cm (6–12 in) from the ground level base of the rootstock.

To make the T-cut, use a sharp knife, preferably a budding knife, and cut horizontally through the bark to make the top of the T. The cut should be about 1 cm ($^3/_8$ in) long, or about one quarter of the circumference of the stock shoot or stem. Next make the vertical cut of the T by inserting the knife about 2.5–6 cm (1–1½ in) below the horizontal cut and pulling the blade upwards until the cuts meet. With luck the bark will spring open, but if it does not, ease it back gently with the back of the knife blade (or the flattened part of the handle if it is a budding knife).

Prepare the bud by holding the bud wood upside down and cutting from below the bud upwards. Just take a sliver of wood, so that there is about 1 cm ($^3/_8$ in) of bark and wood above and below the bud and leaf stalk. With the point of the knife, carefully remove the woody piece, leaving only the bark, bud and leaf stalk shield. Using the leaf stalk to hold the piece the correct way up in one hand, lift back

Propagating sansevieria by means of
leaf cuttings. Cut a leaf into sections
and trim the lower portion of each
into an inverted wide-angled V-shape
so that the main vein makes good
contact with the rooting medium. New
plants will form from the bases.

Azalea occidentalis 'Superba' and other deciduous varieties can be increased by layering at any time of year, grafting in spring on to *Rhododendron luteum* rootstock, semi-hardwood cuttings taken in summer and
put in a cold frame, or from seed.

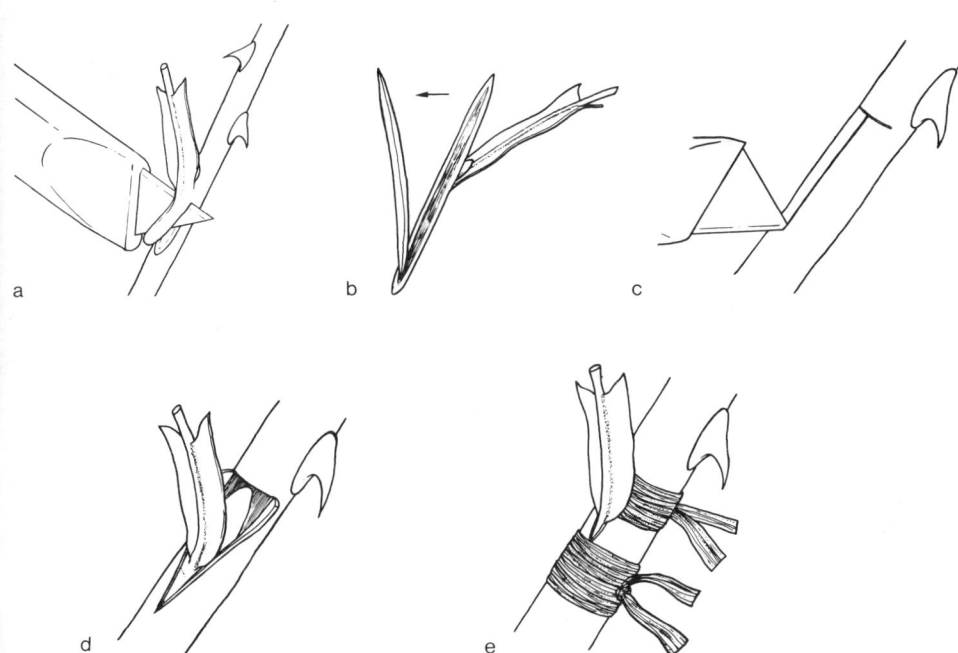

Fig. 15 Budding roses. (a) Cut a bud and its leaf stalk from half way up the chosen stem, cutting upwards with just a sliver of bark and wood. (b) Remove the woody piece, leaving only bark, bud and leaf stalk. (c) Make a T-shaped cut in the bark of the stock. (d) Ease back the bark flaps and slide in the bud. (e) Cut away any of the shield above the T-cut, replace the stock bark, and bind with raffia or proprietary tie above and below the cut surfaces.

the flaps of bark where there is the T-cut with the other, and slide the bud shield down into the vertical section of the cut. Settle it firmly in position and, if there is a bit of the shield sticking up above the vertical part of the T-cut, remove it with a sharp horizontal cut without damaging the bark of the stock plant. Replace the flaps of bark over the bud shield and use a proprietary bud tie or raffia to hold the two firmly together. Make sure the tie starts and finishes below and above the cut surfaces, leaving only the leaf stalk and the bud protruding. In the case of budding, waxing is unnecessary (Fig. 15.)

After several weeks a satisfactory stock and scion union should have taken place. This can be seen by the state of the bud. If it is still plump or beginning to show signs of growth, budding has been successful. If the bud is shrivelled and brown, however, something has gone wrong and it will be necessary to re-bud the stock, starting the whole operation again with fresh stock cuts and new buds.

Remove the bud ties from successful unions after five to six weeks and, in the case of bush roses, gently push back the soil so that it reaches to just below the bud.

The budded plants are left to grow on and are treated normally—weeding, watering, feeding, controlling pests and diseases, etc.—until the following spring. The shoots and stems of the rootstock are then cut away about 1 cm ($^3/_8$ in) above the budded unions and discarded. Any shoots on the rootstock below the union are removed also, as are root suckers. Such unwanted growths must be removed throughout the life of the plant.

It is usual with plants to be grown as bushes, especially roses, to cut out the tip of the new shoot growing from the bud at an early stage, to encourage bushy, well-shooted growths. Single stemmed plants of top fruits, such as cordons, espalier or fan-trained ones, should be given the support of canes or wires at this stage to prevent breakage or damage to the young shoot.

11 Growing from Seed

Although this book is concerned primarily with the raising of new plants by vegetative (asexual) propagation, it would be incomplete not to include a chapter on raising plants from seed (sexual propagation). The end results are what will probably concern most gardeners, and these prove that plants propagated by vegetative methods will produce plants identical to their parents (except, perhaps, in the question of vigour if grafting or budding), whereas plants raised from seeds produce new plants which may sometimes be entirely different from their parents.

Fortunately, many annuals, biennials and vegetables produce seeds that are true to type and raising new plants from them is a cheap and simple method. On the other hand, perennial herbaceous plants, shrubs and trees are not so reliable, and plants from seed originally labelled F_1 hybrids very rarely produce results true to type in the second and subsequent generations. This is due to the necessity of two pure-bred strains of a plant being cross-pollinated under strict cultural conditions to give additional vigour and uniformity of performance.

The pollination process, in order to produce the seeds required, is a complicated and lengthy procedure, usually undertaken only by commercial growers and seed firms. It is, however, quite possible to harvest seeds of many plants from your own garden and raise new plants from them. Alternatively, of course, packets of seeds can be purchased and new plants raised from sowing these. In the latter case, sowing and planting instructions are usually clearly set out on each packet.

As a general rule, hardy annual and vegetable seeds are sown directly outdoors in spring, so also are biennial flowers, but not until summer. Half-hardy annuals, and vegetables required for early cropping, are raised from seeds sown in heat in winter or early spring, for planting out when the danger of frost is over.

HARVESTING SEEDS FROM THE GARDEN

As soon as they are ripe, collect pods, fruits or seed heads from the chosen plant or plants. To obtain enough seed, it may be necessary to harvest more than once from the plants. Collect the seeds, if possible, on a dry day and then lay them on pieces of paper, in saucers or similar flat containers, and carefully label each batch with the names of the plants from which they came. Leave them to finish ripening in a warm, dry, sunny and well-ventilated position for one to two weeks.

When fully dry, clean each batch of seed, by shaking them out of their pods or rubbing them between fingers and thumb. Remove the waste material and, to clean the seeds still further, gently blow on the seeds to remove any dust or chaff still remaining. Put the cleaned seeds in labelled, unsealed packets or envelopes. Store them in a cool, dry but well-ventilated place until sowing time comes around.

The length of time that seeds remain viable (still able to germinate) varies enormously. It may be for only up to one year, or it can be for as much as ten years or longer.

Seeds that are borne in fleshy fruits or berries, or which have very hard coats, often need what is called a period of stratification. This is a period of cold in a moist medium so that the fleshy part rots away and the hard seed coats are softened. This is most usually done by putting the fruits or berries in labelled pots or boxes of a 50:50 mixture of moist peat and sand and placing them outdoors in the garden, covering them with netting to prevent disturbance from birds, rodents or pets. Again the length of time before germination takes place can vary considerably from three or four months up to two years.

If only very few fruits or berries are to be used to supply seeds, it is quite possible to squash them by hand and pick

76

out the individual seeds. These are best sown immediately in a suitable seed sowing compost and placed outdoors and treated as described above. This treatment also applies to very hard coated seeds, when the cold and damp of winter weather helps to break the dormancy period. As an alternative, chipping the hard seed coat with a knife or nail file before sowing in spring will often encourage germination and obviate the need for stratification.

REQUIREMENTS FOR SEED GERMINATION

The three most important requirements for seed germination are moisture, warmth and air. Also important, but usually not so vital, are the time of sowing, the depth at which the seeds are sown, and the medium in which the seeds are to germinate. In most cases, seeds do not require light in order to germinate, though the seedlings will as soon as the first shoots and seed leaves appear.

Moisture is necessary to help promote growth after the dry period of dormancy. Water is supplied artificially by a fine rosed watering can or sprayer to seeds sown in pots or boxes, but is generally available naturally to seeds sown in the open ground (if the soil is dry during drought periods, watering may be necessary). In all cases, however, care must be taken not to over-water seeds, or to cause the soil to become water-logged or form a hard crust on the surface. These defects could cause lack of air which would inhibit germination, or result in a stagnant, moist growing medium in which the seeds could drown and rot.

Warmth is necessary for germination, but the temperature at which seeds germinate best varies considerably according to genus, species and varieties. As a very rough guide, hardy plants will germinate in a temperature of 7°C (45°F) or even less (though germination will be slow); most seeds respond well at about 10–13°C (50–55°F), but some greenhouse subjects and houseplants, particularly those of a tropical nature, may require upwards of 24°C (75°F). Too high a germination temperature can lead to weak seedlings.

The time of sowing is usually dictated by the period at which the adult plant is to mature, flower or fruit. In general, most seeds are sown from early to late spring.

As a general rule, the depth at which to sow seeds should be double that of their thickness; fine seeds should have only a thin layer of seed sowing medium or silver sand scattered over them, larger seeds a reasonable covering, and very large seeds, such as peas and beans, can, in fact, be buried to a depth of 5–7.5 cm (2–3 in). Very fine, almost dust-like seeds are best not covered at all, though protection from light by using paper is usually essential.

The growing medium should be one that does not pack down, is not difficult to water, and which allows the free penetration of air. It is also useful if it contains a little plant food to help sustain healthy growth of the seedlings after they have used up supplies in the seed and before they are ready for transplanting. For greenhouse or home use there are a number of good proprietary seed sowing composts available, some based on soil, such as the John Innes type, and others which are soilless. An inert sterile medium, such as vermiculite, can be used, but the seedlings will need very dilute feeds of liquid fertilizer at an early stage to promote healthy growth. Provided the soil is well cultivated and prepared prior to sowing, seeds sown direct into the open ground should have their necessary requirements met naturally.

SOWING SEEDS OUTDOORS

Many hardy annuals, biennials, herbaceous plants, trees, shrubs and vegetables, also lawns, are raised from seed sowings made direct into the open ground. Sometimes they are sown in seed-beds maintained especially for the purpose, from which the young plants are moved at a later date to their permanent positions, and on other occasions the seeds are actually sown *in situ*, i.e. in their final sites.

Wherever seeds are sown outdoors, the preparation of a good seed bed is of paramount importance. The ground should be prepared well in advance, preferably during late autumn prior to sowing the following spring. Dig the ground thoroughly and incorporate well-rotted manure, compost or similar organic matter in the process. If lime is required, add this to the soil sometime during the winter months, but leave a two to three month period between applications of

78

animal manure and lime; the two together will do more harm than good to the soil. Weather conditions during the winter will help break the soil up into a flexible, workable, fertile and well-aerated condition.

Prior to seed sowing the final preparations of the soil should take place. Lightly fork over the soil, removing weeds and large stones, then rake in a general-purpose fertilizer at the rate of 60 g per sq m (2 oz per sq yd). Continue raking backwards and forwards, and crosswise, until the surface is fine and crumbly. If the soil is heavy and wet, a layer of peat raked in at this stage will assist seed germination. Tread the soil firm by shuffling the feet over the prepared area, then lightly rake it once more. Provided the ground is moist but not too wet it is then ready for seed sowing.

Seeds are sown in the open by two methods: in straight-line drills (furrows), or scattered (broadcast). The latter method is the easiest but generally used only for sowing hardy annual flowers and lawn seed *in situ*. The seed is merely scattered thinly (or at the recommended rate) by hand in the desired position and the soil raked over lightly. If patches of flowering annuals are sown, allow them to overlap slightly so that there are no hard lines of specific groups of plants in the border. Thin the young plants to the distance apart recommended on the seed packet. Sometimes the thinned out annuals can be transplanted satisfactorily to other sites. Watering during the germination period should not be necessary if the soil is moist at the time of sowing. If not, soak it beforehand and leave it for a day or so to dry out a little before sowing the seeds. Watering the seedlings once they are above ground is often necessary and this is best done with a fine spray attached to a can or hose, or even an automatic sprinkler, so that the moisture penetrates through several inches of soil and gets to the root area. Weed as necessary, preferably by hand.

Sowing seeds in drills is the most commonly used method for vegetables sown *in situ* and plants raised in nursery seed-beds, as it is generally more convenient when cultivating the plants after germination; it also tends to ensure the seeds are sown to the correct depth with a suitable covering of soil. The distance apart of each drill will depend on the plants to be grown, and can vary from as little as 15 cm (6 in) to as much as 60 cm (2 ft.).

To make the necessary drill, which should generally be V-shaped, a garden line and hoe are necessary. Occasionally the drill should have a flat bottom—for peas and beans for example which are spaced out in the drill to form a double row. Peg the line firmly at each end of the length of ground where the drill is to be made (for very short drills, a straight piece of wood or a cane laid horizontally on the ground can be used). With the edge of the hoe blade, pull it backwards through the soil against the taut line, so that the trench is at the required depth along the full length. Repeat the process at the necessary distances apart to give the number of drills required. If the soil is at all dry, water the base of the drills to aid germination.

The seed is then sown in the drills as thinly as possible by hand, 'sifting' it out between the fingers and thumb. Large seeds, and those which are pelleted (covered with a clay-soil mixture to make it possible to sow them individually and so save on the quantity of seed and the necessity of thinning), are sown separately to the recommended distance apart. Some gardeners also prefer not to sow a continuous drill of smaller seeds but to place a tiny pinch at intervals at the final distance apart recommended for the mature plant; surplus seedlings are then merely removed leaving one *in situ*.

Having sown the seeds, use the back of a rake to pull the soil over the drills. Firm the soil lightly with the head of the rake and label each row of seeds if necessary. Rake the area lightly to remove footmarks.

Once the seedlings have germinated thin as necessary to the recommended distances. Use the seedlings for further plantings if required. Those raised in seed-beds should be transplanted to their final positions when the time is right. Always try to do any transplanting at an early stage so the young plants receive the minimum check to their growth. Use a dibber to make the planting holes for the small root system and firm the soil lightly over the roots. To prevent wilting, transplant on a dull day and gently water after planting. Larger plants, such as bedding plants, will have a

80

greater root system when transplanted and the hole should be made with a trowel. Again, water after planting. In both cases, if the weather is dry and hot spraying the transplanted plants with water for several days after they have been set in position will help them to become freely established. At all times water, feed, weed and control pests and diseases as necessary to ensure healthy growth of the young plants. (The application of chemical seed dressings at the time of sowing is a wise precaution.)

When sowing seeds in the open, especially in the early part of the year, it is often helpful to warm up the soil beforehand by placing cloches over the areas to be sown. Equally, the protective effect of cloches after sowing is beneficial in providing earlier crops, particularly of vegetables. Another method of obtaining earlier results is to sow seeds in a cold frame and then plant out the seedlings, but this involves more labour.

SOWING SEEDS IN HEAT
Those plants which are only half-hardy or are for growing in the greenhouse or home, or where an early crop yield of certain vegetables, such as onions, leeks, brassicas and lettuce, are required, and all of which are to be propagated from seed, will require the benefit of heat early in the season—late winter and early spring—to enable germination to take place. All but the tender plants can finally be hardened off sufficiently for planting outdoors when the danger of frost is past.

If you have a heated greenhouse, or a part of it is heated, and the root temperature area can be maintained at 10–13°C (50–55°F), this will be adequate for most plants. As mentioned previously, higher temperatures will be required to germinate seeds of tropical plants. If no greenhouse is available, a proprietary heated propagating case can be used, or a warm room in the home. In all instances, the procedures for sowing the seeds and caring for the seedlings follow the same pattern.

Hygiene is of considerable importance as the warm, moist environment is conducive to the spread of pests and diseases. Therefore make sure that all equipment to be used is thoroughly cleaned in advance, by washing in disinfectant if necessary (and this includes the greenhouse itself). Also use only specially prepared composts, such as the proprietary John Innes seed and potting composts or the equivalent soilless ones. Damping off of seedlings is often a problem and watering with Cheshunt Compound will do much to prevent this disease, though any infected seedlings should be removed promptly and be destroyed.

Fill the pots or boxes to be used with a seed sowing compost to within 2 cm (¾ in) of the top and press it down firmly with the fingers, or a piece of flat wood used as a tamper, ensuring the surface is level. Water lightly with a fine rosed can and leave overnight. The following day, sow the seeds as thinly as possible, spacing out the larger ones individually 5 cm (2 in) apart (or placing them separately in small pots, peat or soil/peat blocks), and scattering the smaller ones between finger and thumb; very small seeds can be 'bulked up' with silver sand to help one to sow thinly. Then scatter a layer of compost or silver sand over the seeds, to a depth which is double the width of the seeds. Using a sieve helps to make this job easier. Gently firm the surface again and give a fine sprinkling of water. Label each pot or box with the name of the plant sown.

Place the pots or boxes of seed in a warm place where the required temperature can be maintained. This may be on the benches of a heated greenhouse, in a warm corner of the greenhouse specially created for the purpose, in a propagating case, or in a warm room in the home. (See Chapter 2 for information about the equipment available and the methods of creating situations with the correct environment in the greenhouse.) Unless in a propagating case, cover the pots and boxes with glass, or polythene, to preserve a moist atmosphere; in the home, placing pots in polythene bags tied at the neck will serve the purpose well. To prevent drops of condensation falling on the compost, wipe the glass or polythene dry daily or open the bags for about half an hour. As most seeds do not require light for germination, a covering of paper laid over the glass or polythene, to keep them in the dark, is beneficial as it reduces the rate at which the compost dries out. Once the seedlings have emerged (usually any time between one and four weeks) remove the paper and give them full light, though avoid direct hot sunlight. Gradually increase the amount of ventilation by lifting the glass or polythene, until after about seven days when it can be removed entirely. During the period from the sowing of the seeds to the appearance of the seedlings, watering is usually unnecessary. Should it be required for any reason, give only a very gentle sprinkling.

In the case of the seeds of some vegetables and half-hardy annuals, they can be purchased already sown in containers and all that has to be done is to water the compost thoroughly and place over the top the plastic 'lid' supplied, to make a miniature propagating case. These containers can be stood anywhere in a warm place until germination takes place.

The earlier the seedlings are transplanted the better, as there is less check to growth and it prevents overcrowding, which could cause weak, lanky plants. Most flower and vegetable seedlings are best transplanted into boxes containing a mild compost, such as John Innes no. 1, or an equivalent soilless mixture. In the case of houseplants, these are best transplanted into small pots containing similar compost, and gradually potted on into larger pots of stronger compost until the final pot size is reached. Where only a few seedlings are to be transplanted, or perennials and biennials are not to be put into their final positions until autumn, some people like to place these direct into peat pots or soil/peat blocks so that there is little or no growth set-back when the plants are put out in their final positions.

Transplanting of each seedling is best done by holding the leaves, not the stem, with the fingers and gently levering the roots out of the compost with a plant label or similar object. Make a hole in the potting-on compost with a dibber or pencil and insert the roots. Gently firm back the soil. Space the seedlings about 5 cm (2 in) apart each way in rows, or 7.5 cm (3 in) for vigorous growers, and when the job is completed water the compost and the seedlings with a fine rosed can. Keep the transplanted seedlings at the same initial temperature and place the containers in a shady place for a few days, until growth recommences. Then stand them in direct light once more.

Gradually ventilation can be increased and heat reduced, though care should be taken to see that greenhouse subjects and houseplants are kept in the correct environment required for normal healthy adult growth. Water regularly and apply weak doses of a proprietary liquid fertilizer if this seems necessary.

Containers of hardy and half-hardy plants are finally hardened-off by moving them to a cold frame or placing them under cloches in a sheltered warm position outdoors. To start with keep the frame lights on and the cloches closed, but gradually increase ventilation until the containers are left open to the weather. Planting out usually takes place when the young plants have been hardened off completely and all danger of spring frosts is over. Space the plants at the recommended distance apart. Biennials and perennials which are not to be planted until later in the year can be set out in a nursery-bed until required, or remain in their pots, but care must be taken to see they are watered regularly and that the compost never dries out.

12 Increasing Hardy Perennial Flowers

Hardy perennial flowering plants can be propagated in a number of ways and an alphabetical list of popular plants and the methods of increasing them are listed below. Half-hardy plants, also annuals and biennials, are mentioned in Chapter 11, though the propagation of the popular chrysanthemums, dahlias and pelargoniums are included at the end of this chapter, and the means of increasing stocks of 'bulbous' plants is covered in Chapter 4. Details of how to carry out the specified methods of propagation will also be found in the relevant preceding chapters. Unless otherwise stated, all cuttings are of the nodal type, and if heat is required 13°C (55°F) should be sufficient.

Acanthus (bear's breeches). Division in autumn or spring; root cuttings in cold frame in spring; seed in cold frame in spring.

Achillea (yarrow). Division in autumn or spring; seed in cold frame in spring.

Aconitum (monkshood). Division in autumn or spring; seed in cold frame in spring.

Agapanthus (African lily). Division in late spring.

Alchemilla (lady's mantle). Division autumn or spring; seed in cold frame in spring.

Alstroemeria (Peruvian lily). Division in spring; seed in cold frame in spring.

Anaphalis (pearl everlasting). Division between autumn and spring; softwood basal cuttings in cold frame in spring; seed in cold frame in spring.

Anchusa (bugloss). Division in spring; root cuttings in cold frame in spring.

Anemone (wind flower). Division in autumn or spring; root cuttings in cold frame in spring.

Anthemis (chamomile). Division in autumn or spring; softwood basal cuttings in cold frame in late spring.

Aquilegia (columbine). Division in autumn or spring; seed in the open in late spring.

Artemisia (wormwood). Division in autumn or spring.

Aster (Michaelmas daisy). Division in autumn or spring.

Astilbe (goat's beard). Division in early spring.

Bergenia. Division after flowering or in autumn.

Campanula (bell flower). Division in autumn or spring; softwood basal cuttings in cold frame in late spring; seed in a cold frame in spring.

Centaurea. Division between autumn and spring.

Centranthus (valerian). Softwood basal cuttings in cold frame in spring; seed *in situ* in late spring or early summer.

Chrysanthemum (shasta daisy). Division in spring; softwood basal cuttings in cold frame in spring. (See end of chapter for half-hardy chrysanthemums.)

Convallaria (lily-of-the-valley). Division autumn to spring.

Coreopsis (tickseed). Division in autumn or spring; seed in cold frame in spring.

Cortaderia (pampas grass). Division in spring.

Cynoglossum. Division autumn to spring; seed in cold frame in spring.

Delphinium. Division autumn to spring; softwood basal heel cuttings in cold frame in late spring; seed in open in early summer.

Dianthus (border carnations and pinks). Layer side shoots or take pipings in summer; seed in cold frame in spring.

Dicentra (bleeding heart). Division in spring; root cuttings in cold frame in spring; seed in heat in spring.

Digitalis (foxglove). Division autumn or spring; seed *in situ* in early summer.

Doronicum (leopard's bane). Division in autumn or spring.

Echinops (globe thistle). Division in autumn or spring; root cuttings in cold frame in winter.

86

Erigeron (flea bane). Division autumn to spring; softwood basal cuttings in cold frame in spring.

Eryngium (sea holly). Division autumn to spring; root cuttings in cold frame in spring.

Euphorbia. Division autumn to spring; softwood basal cuttings in cold frame in late spring; seed in cold frame in spring.

Filipendula. Division autumn to spring.

Gaillardia (blanket flower). Division in spring; semi-hardwood cuttings in late summer, over-wintered in cold frame; seed in cold frame in spring.

Geranium (crane's bill). Division in autumn or spring; seed in cold frame in spring.

Geum (avens). Division in spring; seed in cold frame in spring.

Gypsophila. Softwood basal cuttings in cold frame in late spring; semi-hardwood cuttings in mid-summer in cold frame; seed in cold frame in spring.

Helenium (sneezewort). Division autumn to spring; softwood basal cuttings in cold frame in spring.

Helianthus (sunflower). Division autumn to spring; softwood basal cuttings in cold frame in spring.

Heliopsis. Division autumn to spring.

Helleborus (Christmas and Lenten roses). Division in late spring after flowering; seed in cold frame in early summer.

Hemerocallis (day lily). Division in autumn or spring.

Heuchera (coral flower). Division in autumn or spring; seed in cold frame in spring.

Hosta (plantain lily). Division in spring.

Incarvillea. Division in autumn; seed in nursery-bed in late spring (takes 3 years to flower).

Iris. Division after flowering or in autumn or spring.

Kniphofia (red-hot poker). Division in spring; seed in cold frame in spring.

Lamium (dead nettle). Division autumn to spring.

Liatris (gay feather, blazing star). Division of roots in early spring; seed in cold frame in spring.

Ligularia. Division in spring.

Limonium (sea lavender). Division of roots in early spring; seed in cold frame in spring.

Linum (flax). Softwood basal cuttings in cold frame in spring; seed in cold frame in spring.

Lupinus (lupin). Softwood basal heel cuttings in cold frame in spring; seed in heat in early spring, or cold frame in late spring.

Lychnis (campion). Division in autumn or spring; seed in cold frame in spring.

Lysimachia. Division in autumn or spring.

Lythrum (loosestrife). Division in autumn or spring.

Malva (mallow). Softwood basal cuttings in cold frame in spring; seed in cold frame in spring.

Monarda (bergamot). Division in spring; softwood basal cuttings in cold frame in spring.

Nepeta (catmint). Division autumn or spring; softwood basal cuttings in cold frame in spring.

Oenothera (evening primrose). Division in spring; seed in cold frame in spring.

Omphalodes (navelwort). Division in spring or mid-summer after flowering.

Paeonia (paeony). Division in autumn, ensuring each piece has roots and buds; seed in cold frame in autumn.

Papaver (poppy). Division in spring; root cuttings in cold frame in winter; seed in cold frame in spring.

Penstemon. Non-flowering side shoot cuttings in cold frame in late summer; seed in heat in late winter.

Phlox. Division autumn or spring; softwood basal cuttings in cold frame in spring; seed in cold frame in spring.

Physalis (Chinese lantern). Division autumn or spring.

Physostegia (obedient plant). Division in autumn or spring; seed in cold frame in spring.

Platycodon (balloon flower). Division in spring; seed in open in late spring.

Polemonium (Jacob's ladder). Division in autumn or spring; seed in cold frame in spring.

Polygonatum (Solomon's seal). Division in autumn or spring.

Polygonum (knotweed). Division in autumn or spring.

Potentilla (cinquefoil). Division in autumn or spring; seed sown in open in spring.

Primula (primrose). Division after flowering or in spring; seed as soon as ripe in cold frame in summer.

Leaf-bud cuttings are best used to increase stocks of *Camellia japonica*. A leaf and leaf stalk are cut from the parent plant with a small piece of stem attached which bears a growth bud, and each cutting is inserted in a suitable rooting medium.

Ericas (heathers and heaths) are propagated from semi-hardwood heel cuttings which root most freely in a mist propagating unit. Layering or division are other means of increasing these plants.

Begonia rex leaves, cut into sections, laid on rooting compost, and which now have new plantlets growing up from the main vein area. They will shortly be ready for planting in individual pots.

Aphelandra squarrosa (zebra plant) is usually increased by softwood cuttings taken in late spring. Alternatively, leaf cuttings may be taken, each leaf with a basal bud and a piece of stem, as here. By cutting off the leaf tip, moisture loss is reduced.

Prunella (self-heal). Division autumn to spring.
Pulmonaria (lungwort). Division after flowering or in
    autumn or spring.
Pyrethrum. Division after flowering or in autumn or
    spring.
Ranunculus (buttercup). Division in autumn or spring.
Rodgersia. Division in spring.
Rudbeckia (cone flower). Division in autumn or spring;
    softwood basal cuttings in cold frame in spring; seed in
    cold frame in spring.
Saponaria. Division autumn to spring.
Saxifraga (saxifrage). Division in spring.
Scabiosa (scabious). Division in spring.
Schizostylis (Kaffir lily). Division in spring.
Sedum (stonecrop). Division in spring.
Sidalcea. Division in autumn or spring.
Solidago (golden rod). Division autumn to spring.
Stachys. Division autumn to spring.
Stokesia. Division in spring; seed in cold frame in spring.
Symphytum. Division autumn to spring.
Tellima. Division autumn to spring.
Thalictrum (meadow rue). Division in spring; seed in cold
    frame in spring.
Tiarella (foam flower). Division in autumn or spring.
Tradescantia (spiderwort). Division in spring.
Trillium (wood lily). Division after leaves die down in late
    summer; seed as soon as ripe in cold frame (germination
    very slow and plants can take six years to flower).
Trollius (globe flower). Division in autumn or spring;
    seeds in cold frame autumn to spring.
Verbascum (mullein). Root cuttings in cold frame in
    spring; seed in cold frame in spring.
Veronica (speedwell). Division in autumn or spring.
Viola (pansy). Non-flowering basal cuttings in cold frame
    in mid-summer; seed in open or cold frame in mid-
    summer.

Many people consider their garden incomplete without
the inclusion of chrysanthemums, dahlias and pelargoniums.
In completely frost-free areas, these plants can be left in the

open all year round but, because such a situation is the exception rather than the rule, they are generally treated as half-hardy plants and are lifted in late autumn and given the protection of a slightly warm storage place in winter. Propagation takes place in the warm, and the new plants are set out each year when the danger of frosts is past.

Chrysanthemums. Most perennial chrysanthemums are raised from softwood basal nodal cuttings in spring in a heated environment. Many varieties to go outdoors are propagated in mid-spring; large flowered types a month or so earlier; early ones a month or so later; and late flowering indoor varieties in early summer.

Suitable growths for cuttings are produced from the 'stool'—the mass of roots and 15 cm (6 in) or so of stems left after the plants have been cut back when they have finished flowering. These 'stools' are washed and packed in boxes of damp peat and kept in a slightly warm place until started into growth by watering and a slight increase of heat about three weeks before cuttings are required. Take softwood basal cuttings, or stem cuttings if the variety does not produce 'stool' ones which are about 5 cm (2 in) long, either by breaking off the young shoots or cutting them just below a node. Remove the lowest leaves and insert the bottom 2.5 cm (1 in) of the cuttings, 5 cm (2 in) apart each way, in pots or boxes of a 50:50 mixture of peat and coarse sand. Water to settle in the cuttings and place the containers in a propagator, polythene bags or cover with glass. A bottom heat of about 15°C (59°F) helps quicker rooting. Sprays of water may be needed from time to time to prevent the cuttings wilting.

Once rooted, transfer the young plants individually into 7.5 cm (3 in) clay, plastic or peat pots, filled with John Innes no. 1 compost or similar, or peat/soil blocks. Keep them in warm conditions until mid-spring at least, then gradually harden off the ones to be grown outdoors by moving them to a cold frame, closed at first, then with the light removed. Plant out when all danger of frosts is past. Varieties that are to be grown in pots, should be potted on into successively larger pots at intervals with stronger compost each time until they reach flowering pot stage, usually the 20–25 cm (8–10 in) size.

Some chrysanthemum varieties can be raised from seed sown in heat, 13–15°C (55–59°F), in late winter. Transplant and grow the seedlings on as recommended for cuttings.

Dahlias. These plants are propagated by division of the tubers, the simplest method as no heat is required, or from basal nodal stem cuttings. In either case the tubers are lifted in late autumn, the stems cut back, the tubers washed clean of soil, dried, and then stored in boxes of peat in a frost-free place.

If increasing by division, in late spring water the peat, not the tubers, and in two to three weeks the 'eyes' (buds) at the bases of the stems will begin to enlarge. Take the tubers out of the peat and, with a sharp knife, cut the stems into sections, making sure each piece is complete with at least one 'eye'. Either plant the divided pieces at least 10 cm (4 in) deep, where they are to flower, or place them in pots or boxes of John Innes potting compost no. 2 in a cold frame until spring frosts are past, when they can be set outdoors.

When propagating dahlias by cuttings, start the tubers into growth as described, and take 7.5 cm (3 in) long soft-wood basal nodal cuttings – shoots produced from the 'eyes' at the bases of the stems. Cultivate these cuttings just as you would those of chrysanthemums and set out the young plants in the open when danger of spring frosts is over.

Dahlias can also be raised from seed. Sow the seed in a temperature of about 15°C (59°F) in late winter to early spring, transplant seedlings to pots and gradually harden off by the usual methods before planting out.

Pelargoniums (geraniums). Cuttings are the best method of propagating these plants. Take 7.5–10 cm (3–4 in) long, non-flowering tip cuttings in late summer or early autumn. Trim these to immediately below a node, remove the lowest leaves and small 'leaflets', and insert the bottom half of each stem into pots or boxes containing a cuttings compost or a 50:50 peat and coarse sand mixture (Fig. 16). Stand these on the greenhouse bench or in a garden frame where there

Fig. 16 Cuttings of pelargoniums. *Left*: trim a non-flowering tip cutting below a node and remove the lowest leaves and small 'leaflets'. *Right*: insert the lower half of the cutting in a suitable compost in a pot.

is just sufficient heat to keep them free from frost, water in the cuttings and spray as necessary until rooting has occurred. Pot up the young plants into 7.5 cm (3 in) pots of John Innes potting compost no. 1 or similar when rooted and keep it moist throughout the winter. Pinch out the growing tips when the plants are 15 cm (6 in) high to encourage bushiness. Harden off and plant out in late spring or early summer.

Pelargoniums lifted in autumn, potted into John Innes no. 2 compost or its equivalent, and kept in a frost-free greenhouse, frame, or similar situation, can have softwood cuttings taken in a similar manner in early spring.

These plants can also be raised from seed sown in late winter in a temperature of about 16°C (61°F). Transplant, pot up, grow and harden off as is usual for seedlings, and set out the young plants when danger of spring frost is over.

13 Propagating Decorative Shrubs, Trees and Climbers

Most shrubs, trees and climbers are increased by cuttings taken from suitable stems or shoots and rooted as described in Chapter 6. Mist propagation is particularly valuable in rooting cuttings of these plants. Where bottom heat is recommended the temperature should be about 15°C (59°F). Other methods of propagation used are also covered in earlier chapters. Seed is often used as a means of increasing stocks and, if so, is generally best sown as soon as ripe in pots or boxes of a 50:50 mixture of sand and peat, with the containers either left in the open, to stratify fleshy seeds during winter and encourage spring germination, or given the protection of a cold frame.

This plant list shows the best propagation methods (cuttings are of the nodal type unless specified otherwise).

Abelia. Semi-hardwood heel cuttings in mid-summer with bottom heat; hardwood cuttings in late autumn in cold frame.

Acer (maple). Graft in spring on to type species; seed in autumn outdoors.

Actinidia. Semi-hardwood cuttings in mid-summer with bottom heat; seed in autumn in cold frame.

Amelanchier (snowy mespilus, June berry). Layer in autumn; take rooted suckers autumn to spring; seed in summer in cold frame.

Arbutus (strawberry tree). Semi-hardwood heel cuttings in mid-summer with bottom heat; layer in spring; seed in autumn in cold frame.

Aucuba (spotted laurel). Semi-hardwood heel cuttings in late summer in cold frame; layer in spring.

Azalea – see Rhododendron.

Berberis (barberry). Semi-hardwood cuttings in late summer in cold frame; division of suckering types autumn to spring; seed in early winter outdoors.

Buddleia (butterfly bush). Semi-hardwood heel cuttings in summer outdoors.

Buxus (box). Semi-hardwood cuttings in late summer in cold frame; division in autumn or spring; layer in autumn.

Calluna (ling)—see Erica.

Camellia. Semi-hardwood cuttings or leaf-bud cuttings in summer with bottom heat; layer in early autumn; seed in early spring in heat.

Caryopteris. Semi-hardwood heel cuttings in late summer in cold frame.

Catalpa (Indian bean tree). Semi-hardwood heel cuttings in mid-summer in heat.

Ceanothus (Californian lilac). Semi-hardwood heel cuttings in mid-summer with bottom heat; hardwood cuttings of evergreen species in late autumn in cold frame.

Chaenomeles (japonica, Japanese quince). Semi-hardwood heel cuttings in mid-summer with bottom heat; layer in summer; seed in autumn in cold frame.

Chamaecyparis (false cypress). Semi-hardwood heel cuttings in late spring in cold frame; softwood heel cuttings in winter in heat; seed in mid-spring outdoors.

Chimonanthus (winter sweet). Layer in early autumn; seed in autumn in cold frame.

Choisya (Mexican orange blossom). Semi-hardwood heel cuttings in summer with bottom heat.

Cistus (rock rose, sun rose). Semi-hardwood heel cuttings in summer with bottom heat; seed in spring in cold frame.

Clematis (virgin's bower). Semi-hardwood internodal cuttings in mid-summer with bottom heat; serpentine layering in spring; seed in autumn in cold frame.

Cornus (dogwood, cornel). Semi-hardwood heel cuttings in summer in cold frame; layer in early autumn; remove self-rooted suckers in late autumn.

Corylopsis. Layer in autumn; semi-hardwood cuttings in summer with bottom heat.

Cotinus (smoke tree). Layer in autumn; semi-hardwood cuttings in summer in cold frame.

Cotoneaster. Semi-hardwood heel cuttings in summer of deciduous species, hardwood heel cuttings in autumn of evergreen species, both in cold frame; layer in autumn or spring; seed in autumn outdoors.

Cryptomeria (Japanese cedar). Semi-hardwood cuttings in late summer in cold frame; seed in mid-spring outdoors.

× *Cupressocyparis* (Leyland cypress). Semi-hardwood cuttings in late summer in cold frame.

Cytisus (broom). Semi-hardwood heel cuttings in late summer in cold frame; seed in mid-spring in cold frame.

Daboecia (Irish bell heather, St. Dabeoc's heath)—see Erica.

Daphne. Semi-hardwood heel cuttings in summer in cold frame; layer evergreen species in spring; seed in early autumn in cold frame.

Deutzia. Semi-hardwood cuttings in mid-summer in cold frame; hardwood cuttings in late autumn outdoors.

Elaeagnus. Semi-hardwood cuttings of evergreen species in late summer in cold frame; sow seed of deciduous species in summer in cold frame.

Enkianthus. Layer in autumn or spring; semi-hardwood heel cuttings in late summer in cold frame.

Erica (heather, heath). Semi-hardwood heel cuttings in summer with bottom heat or in mist propagating unit; layer in spring; division in spring.

Escallonia. Semi-hardwood heel cuttings in summer in cold frame or with bottom heat.

Euonymus (spindle tree). Semi-hardwood lateral shoot heel cuttings in late summer in cold frame; seed in early autumn in cold frame.

Fatsia (castor oil plant). Root detached sucker shoots in mid-spring in cold frame; seed in spring in heat.

Forsythia (golden bell bush). Hardwood cuttings in mid-autumn outdoors; layer in mid-autumn; remove self-rooted layers in autumn.

Fothergilla. Layer autumn or spring.

Fuchsia. Softwood non-flowering tip cuttings from spring to autumn with bottom heat; seed in spring in heat.

Garrya (silk tassel bush). Layer in autumn; semi-hardwood heel cuttings in late summer in cold frame.

Gaultheria. Layer or detach rooted suckers in early autumn; semi-hardwood heel cuttings in mid-summer in cold frame; seed in autumn in cold frame.

Genista (broom). Semi-hardwood heel cuttings in late summer in cold frame; seed in spring in cold frame.

Hamamelis (witch hazel). Layer in early autumn.

Hebe (veronica). Semi-hardwood cuttings in summer in cold frame.

Hedera (ivy). Semi-hardwood cuttings (runner growths for climbers—adult growths for bushy forms) in summer with bottom heat or in mist propagating unit; hardwood cuttings in early winter outdoors; serpentine layering in spring.

Helianthemum (rock rose, sun rose). Semi-hardwood cuttings in summer in cold frame.

Hibiscus (tree mallow, tree hollyhock). Semi-hardwood heel cuttings in early summer with bottom heat.

Hippophaë (sea buckthorn). Layer in autumn; root cuttings in spring outdoors; seed in early autumn in cold frame.

Hydrangea. Semi-hardwood cuttings of bushy species in late summer with bottom heat or in cold frame; semi-hardwood cuttings of climbing species in early summer in cold frame.

Hypericum (St. John's wort, rose of Sharon). Division autumn to spring; softwood cuttings of small species in late spring, semi-hardwood cuttings of tall species in summer, both in cold frame.

Ilex (holly). Semi-hardwood heel cuttings in mid-summer in cold frame; layer species and their forms in autumn.

Jasminum (jasmine). Layer in early autumn; semi-hardwood heel cuttings in mid-summer, or hardwood heel cuttings in early winter, both in cold frame; seed in early autumn in cold frame.

Juniperus (juniper). Semi-hardwood heel cuttings in early autumn in cold frame; seed in early autumn in cold frame.

An easy method of increasing streptocarpus is by taking leaf cuttings in early summer and cutting an inverted V-shape at the base before inserting in a rooting compost. New plants arise from the base of the main vein.

Young pelargonium (geranium) plants which have been raised from non-flowering tip cuttings taken in early autumn.

Fuchsias, such as this 'Eva Borg' variety, are increased by taking non-flowering tip cuttings any time from spring to autumn, and rooting them in a position where they receive bottom heat.

Kalmia (sheep laurel, calico bush). Layer in late summer; semi-hardwood cuttings in summer in cold frame.

Kerria (Jew's mallow). Division autumn to spring; semi-hardwood cuttings in late summer outdoors.

Kolkwitzia. Semi-hardwood heel cuttings in early summer in cold frame.

Laburnum (golden rain, golden chain). Graft in spring on species; seed in autumn in cold frame.

Laurus (sweet bay, bay laurel). Semi-hardwood cuttings in late summer in cold frame; layer in summer.

Lavandula (lavender). Semi-hardwood cuttings in summer in cold frame; hardwood cuttings in late autumn outdoors.

Leycesteria. Hardwood cuttings in autumn outdoors; semi-hardwood cuttings in summer in cold frame; seed in spring in cold frame.

Ligustrum (privet). Hardwood cuttings in late autumn outdoors.

Lonicera (honeysuckle). Semi-hardwood cuttings in summer in cold frame; hardwood cuttings in autumn in open; serpentine layering in late summer; seed of climbing species in autumn in cold frame.

Magnolia (tulip tree). Layer in mid-spring; air layer in summer; semi-hardwood cuttings in mid-summer with bottom heat; seed in autumn in cold frame.

Mahonia. Semi-hardwood cuttings in early summer with bottom heat; rooted suckers in autumn or spring; seed in late summer in cold frame.

Malus (crab apple, flowering crab). Graft in spring or bud in summer on selected rootstocks.

Olearia (daisy bush). Semi-hardwood heel cuttings in summer in cold frame.

Osmanthus. Layer in autumn; semi-hardwood cuttings in summer with bottom heat; hardwood cuttings in autumn in cold frame.

Paeonia (tree paeony). Layer in spring; hardwood heel cuttings in autumn in cold frame; seed in autumn in cold frame.

Parthenocissus (Virginia creeper). Layer in late autumn; semi-hardwood cuttings in late summer with bottom heat; hardwood cuttings in early winter outdoors.

Pernettya (prickly heath). Division in spring; layer in autumn; semi-hardwood heel cuttings in late summer in cold frame; seed of species in autumn in cold frame.

Philadelphus (mock orange). Hardwood cuttings in late autumn outdoors; semi-hardwood cuttings in summer in cold frame.

Phlomis (Jerusalem sage). Division in autumn or spring; semi-hardwood cuttings in late summer in cold frame; seed in late spring in heat.

Picea (spruce, fir). Some species by softwood cuttings in summer in cold frame or mist propagating unit; all by seed in spring in cold frame.

Pieris (lily-of-the-valley tree). Layer in autumn; air layer in spring; semi-hardwood heel cuttings in summer in cold frame; seed in winter in cold frame.

Pinus (pine). Graft varieties on type species in spring outdoors; seed in spring in cold frame.

Polygonum (Russian vine, mile a minute vine). Hardwood cuttings in early winter outdoors; semi-hardwood cuttings in cold frame in summer.

Populus (poplar). Suckers in autumn or early spring outdoors; hardwood cuttings in autumn outdoors.

Potentilla (cinquefoil). Semi-hardwood heel cuttings in summer in cold frame; hardwood cuttings in late autumn in cold frame; seed in spring in cold frame.

Prunus (ornamental almonds, apricots, cherries, peaches, plums, cherry laurels). Graft in spring or bud in summer on *P. avium* rootstock; hardwood cuttings of evergreen types in autumn in cold frame; seed of species in autumn outdoors.

Pyracantha (firethorn). Semi-hardwood heel cuttings in summer in cold frame; seed in autumn in cold frame.

Pyrus (flowering pear). Graft in spring or bud in summer on selected rootstock; hardwood cuttings of evergreen types in autumn in cold frame; seed of species in autumn outdoors.

Rhododendron (including azalea). Layer any time of year; graft in spring on *R. ponticum* (or *R. luteum* for deciduous azaleas) rootstocks; semi-hardwood cuttings in summer in cold frame; seed in spring in heat.

Rhus (sumach). Rooted suckers in autumn or spring; root cuttings in winter in cold frame; layer autumn or spring.

Ribes (flowering currant). Hardwood cuttings in late autumn outdoors.

Robinia (false acacia). Remove rooted suckers in autumn or spring; seed in spring in cold frame.

Rosa (rose). Budding in summer on selected rootstock; hardwood cuttings in autumn outdoors (not hybrid teas); seed in autumn outdoors.

Rosmarinus (rosemary). Semi-hardwood cuttings in summer in cold frame; hardwood cuttings late autumn or spring outdoors.

Rubus (ornamental brambles). Division autumn to spring; tip layer in autumn; semi-hardwood heel cuttings in summer in cold frame.

Ruscus (butcher's broom). Division autumn to spring.

Salix (willow). Hardwood cuttings in winter outdoors.

Sambucus (elder). Hardwood cuttings in winter outdoors; semi-hardwood cuttings in summer in cold frame.

Santolina (cotton lavender). Semi-hardwood cuttings in summer in cold frame.

Sarcococca. Division autumn or spring; semi-hardwood cuttings in late summer in cold frame.

Senecio. Semi-hardwood cuttings in late summer in cold frame.

Skimmia. Layer in autumn; semi-hardwood heel cuttings in summer in cold frame.

Sorbaria. Rooted suckers autumn to spring; hardwood cuttings late autumn outdoors.

Spiraea. Division or rooted suckers autumn to spring; semi-hardwood cuttings in summer in cold frame; hardwood cuttings in late autumn outdoors.

Stuartia. Layer in autumn; semi-hardwood heel cuttings in mid-summer in cold frame; seed in autumn in cold frame.

Symphoricarpos (snowberry). Rooted suckers or division autumn to spring; hardwood cuttings late autumn outdoors.

Syringa (lilac). Semi-hardwood shoots in summer with bottom heat; layer in spring; graft in spring or bud in winter *S. vulgaris* varieties on type species or ligustrum (privet).

Tamarix (tamarisk). Hardwood cuttings in late autumn outdoors.

Taxus (yew). Semi-hardwood heel cuttings in late summer in cold frame; seed in late autumn in cold frame.

Thuja (arbor-vitae). Semi-hardwood heel cuttings in late autumn in cold frame; seed in spring in cold frame.

Ulex (gorse, whin, furze). Semi-hardwood cuttings in late summer in cold frame; seed in late spring in cold frame.

Viburnum. Layer in autumn; semi-hardwood heel cuttings in early summer with bottom heat or in late summer in cold frame; seed in autumn in cold frame.

Vinca (periwinkle). Division autumn to spring; layer in autumn.

Vitis (ornamental vine). Layer in autumn; hardwood cuttings in early winter outdoors; semi-hardwood heel or eye (bud) cuttings in summer.

Weigela. Hardwood cuttings in late autumn outdoors; semi-hardwood heel cuttings in early summer with bottom heat.

Wisteria. Layer in autumn or spring; semi-hardwood cuttings in mid-summer with bottom heat or in mist propagating unit; graft on *W. sinensis* roots in spring in heat.

Yucca. Remove rooted suckers in spring.

14 Greenhouse and Houseplant Propagation

Plants which are not hardy enough to be grown in the open garden all the year round, and which require the protection of a warm environment for all or part of the year, are generally referred to as greenhouse plants or houseplants. The temperatures and conditions they require when mature will vary according to the type of plant.

Increasing stocks of such plants is carried out by a variety of different forms of cuttings, or the sowing of seed, but all need heated conditions and, for best results, the equipment described in Chapters 1 and 2 will be of considerable assistance. The average temperatures required for propagation are usually between 13°C (55°F) and 18°C (64°F), but may be as high as 24°C (75°F) and if so, are given as such. Details of the propagation methods applicable are covered in the relevant preceding chapters. The young plants are transplanted and potted on into pots of varying sizes containing increasing strengths of compost until they reach the final pot size necessary for the degree of maturity required. They are watered, fed, pruned and treated against pest and disease attacks as necessary.

Listed below are some of the more popular greenhouse and room plants, and ways of propagating them:

Abutilon (Indian mallow). Semi-hardwood cuttings in
    summer; seed in late spring.
Acalypha hispida (copper leaf). Softwood cuttings in
    spring at 24°C (75°F).

Achimenes (hot-water plant). Division of rhizomes in winter; softwood and leaf stalk cuttings in spring; seed in early spring at 21°C (70°F).

Aechmea fasciata. Suckers in spring.

Agave (American aloe). Offsets in spring and early summer.

Anthurium (flamingo plant, tail flower). Division of roots in early spring; seed in spring at 24°C (75°F).

Aphelandra squarrosa (zebra plant). Softwood cuttings in late spring.

Ardisia crispa (spear flower). Softwood cuttings in summer; seed in spring at 21°C (70°F).

Asparagus myersii (asparagus ferns). Division in spring; seed in spring.

Aspidistra. Division of roots in early spring.

Azalea (Rhododendron simsii). Semi-hardwood cuttings in early summer, preferably in mist propagating unit; layer in spring.

Begonia. Division of tubers of tuberous types in spring; leaf section cuttings of ornamental leaved sorts in early summer; softwood cuttings of fibrous-rooted and winter flowering kinds in late spring; seeds of all types in spring.

Beloperone guttata. Softwood cuttings in early summer.

Billbergia nutans. Division of roots, or removal of suckers in early summer.

Bougainvillea. Softwood cuttings in late spring.

Bouvardia. Division in spring; softwood cuttings in late spring.

Brunfelsia. Softwood cuttings in spring; seed in spring at 21°C (70°F).

Cacti and Succulents. Cuttings in spring; seed in spring.

Caladium. Division of tubers in spring.

Calathea. Division of roots in spring.

Calcéolaria (slipper flower). Shrubby kinds from semi-hardwood cuttings in autumn; other types from seed in early summer.

Campanula. Softwood cuttings in spring; seed in summer.

Canna. Division of rhizomes in spring; seed in spring at 21°C (70°F).

106

Chlorophytum. Division of roots, or layering of shoot tips, in spring.

Chrysanthemum. Basal softwood cuttings in spring or summer.

Cissus antarctica (kangaroo vine). Softwood cuttings in early summer.

Clerodendrum. Softwood cuttings in late spring at 21°C (70°F).

Clianthus puniceus (parrot's bill). Softwood cuttings in early summer; seed in late winter.

Clivia miniata. Division of roots in spring; seed in spring at 21°C (70°F).

Codiaeum (croton). Softwood cuttings in late spring at 24°C (75°F).

Coleus (flame nettle). Softwood cuttings in late spring to mid-summer; seed in spring.

Columnea. Softwood cuttings in late spring.

Cordyline (cabbage palm). Removal of suckers in spring; 7.5 cm (3 in) long softwood tip cutting and cuttings of main stem in early summer.

Crassula. Leaf cuttings or softwood cuttings in early summer; seed in late spring.

Cryptanthus (earth star, star fish). Remove offshoots in late spring.

Cuphea ignea (Mexican cigar flower). Softwood cuttings in spring; seed in early spring.

Cyclamen. Seed in late summer or late winter.

Cyperus (umbrella plant). Division in summer.

Dianthus (perpetual flowering carnation). Side shoot softwood cuttings in late winter.

Dieffenbachia. Remove basal suckers in spring; 7.5 cm (3 in) long softwood tip cutting and cuttings of main stem in early summer at 21°C (70°F).

Dracaena (dragon plant). Basal shoots in spring; air layering in summer; 7.5 cm (3 in) long softwood tip cutting and cuttings of main stem in spring at 21°C (70°F).

Euphorbia pulcherrima (poinsettia). Softwood cuttings in late spring at 21°C (70°F).

Ferns. Division in spring; layering of fronds (leaves) in

early summer; remove bulbils when two or three small fronds appear.

Ficus (rubber plant). Tip cuttings in late spring or early summer; air-layering in summer.

Freesia. Separate offset corms in early autumn; seed autumn to mid-spring.

Fuchsia. Softwood cuttings in spring; seed in spring.

Gardenia. Softwood heel cuttings in spring.

Gloxinia (Sinningia speciosa). Division of tubers, each piece with one or two shoots, in spring; softwood basal shoots from tubers in late spring; leaf cuttings in summer; seed in late winter at 21°C (70°F).

Grevillea robusta (silk oak).Semi-hardwood cuttings in summer; seed in spring.

Hedera (ivy). Tip cuttings in late summer and autumn.

Hippeastrum (amaryllis). Offsets, early spring; seed, spring.

Hoya carnosa (wax flower). Softwood or hardwood cuttings in early summer; layer in late spring.

Hyacinthus (hyacinth). Bulblets in autumn.

Hydrangea. Softwood cuttings in spring.

Impatiens (balsam, busy Lizzie). Softwood cuttings spring to autumn; seed in spring.

Jasminum polyanthum (jasmine). Softwood heel cuttings in spring; layer in autumn.

Kalanchoe. Leaf stem cuttings or softwood cuttings in summer; seed in spring.

Lachenalia (cape cowslip). Divide bulbs in late summer.

Lilium (lily). Division of bulbs, rhizomes, or scales in early autumn; seed in early spring.

Maranta. Division in spring.

Monstera deliciosa (shingle plant). Shoot tip cuttings in early summer; 7.5 cm (3 in) lengths of main stem cuttings in early summer (both at 24–27°C (75–81°F).

Narcissus (daffodil). Division of bulbs after flowering.

Nerine (Guernsey lily). Division of offsets after flowering.

Nerium oleander (oleander). Semi-hardwood cuttings in summer; seed in late spring.

Orchids. Most by division of roots in spring.

Pelargonium (geranium). Cuttings in summer or early spring; seed in late winter.

Peperomia (pepper plant). Softwood stem or leaf stem cuttings late spring to mid-summer.

Philodendron. Softwood 10 cm (4 in) tip cuttings and cuttings of main stem in early summer; division in early summer; seed in late autumn at 24°C (75°F).

Pilea (aluminium plant). Softwood cuttings in early summer.

Plumbago capensis. Softwood basal or side-shoot cuttings in late spring.

Primula. Division after flowering; seed in spring.

Rhoicissus rhomboidea. Softwood cuttings in late spring.

Saintpaulia (African violet). Leaf stem cuttings in summer; seed in early spring.

Sansevieria trifasciata (mother-in-law's tongue). Removal of suckers or by leaf section cuttings, both in summer at 21°C (70°F).

Saxifraga stolonifera (mother of thousands). Remove runner plantlets in summer.

Scindapsus. Tip or softwood basal cuttings in summer at 21°C (70°F).

Solanum capsicastrum (winter cherry). Softwood cuttings in spring; seed in spring.

Sparmannia africana (African hemp). Softwood tip cuttings from cut back plants in late spring.

Spathiphyllum wallissii. Division in spring after flowering.

Strelitzia reginae (bird of paradise flower). Division in spring; seed in spring.

Streptocarpus (cape primrose). Division in spring; leaf cuttings in early summer; seed in spring.

Tolmiea (pig-a-back plant). Remove runner plantlets or layer runners in early summer.

Tradescantia (wandering Jew). Softwood cuttings in summer.

Vallota (Scarborough lily). Remove offsets when re-potting.

Vriesea. Remove offsets in spring.

Zantedeschia (arum lily). Division of rhizomes in early autumn; remove offsets when re-potting.

Zebrina pendula. Softwood cuttings in summer.

15 Propagation of Fruits, Vegetables and Herbs

In the main, all fruits are increased by vegetative means. A large number of vegetables and herbs on the other hand are best raised each year from seed.

FRUITS

Soft fruits, sometimes called berried or bush fruits, are easy to propagate, but tree fruits require a certain amount of expertise and gardeners often leave the production of these to expert nurserymen. Nevertheless, with a little care and practice tree fruits can be propagated in the garden and it is rewarding to obtain good results.

Details of how each fruit is increased vegetatively are given in the previous chapters. Always make sure that whatever part of the plant is used it is free from disease—this applies to the parent plant as well.

Apple (Malus sylvestris). Graft in spring or bud in summer using suitable apple rootstocks.
Apricot (Prunus armeniaca). Graft in spring or bud in summer on plum stocks.
Blackberry, Loganberry (Rubus ursinus, R. loganobaccus). Tip layer in summer; remove rooted suckers in winter.
Black Currant (Ribes nigrum). Hardwood cuttings in autumn outdoors; layer in spring by mounding soil over low-growing shoots.
Cherry (Prunus avium, P. cerasus). Graft in spring or bud in summer on wild cherry rootstocks.

110

Cobnut (Corylus avellana). Layer in autumn; seed in early winter in cold frame.

Currants—red and *white (Ribes sativum)*. Hardwood cuttings in autumn outdoors, removing all but top 3–4 buds to give clean lower stem.

Fig (Ficus carica). Layer in summer; semi-hardwood cuttings in late summer in cold frame.

Gooseberry (Grossularia uva-crispa). Hardwood cuttings in autumn outdoors, removing all but top 3–4 buds to give clean lower stem.

Grape (Vitis vinifera). Eye bud cuttings in early spring with bottom heat (13°C (55°F)); hardwood cuttings in early winter outdoors.

Peach, Nectarine (Prunus persica). Bud in summer on seedling peach or plum stocks; seed in early autumn in cold frame.

Pear (Pyrus communis). Graft in spring or bud in summer on suitable pear or quince stocks.

Plum, Damson (Prunus domestica, P. insititia). Graft in spring or bud in summer on plum rootstocks.

Raspberry (Rubus idaeus). Remove rooted sucker shoots in winter.

Strawberry (Fragaria × ananassa). Layer runners in summer; seed in autumn in cold frame or early spring in bottom heat (13°C (55°F)).

PERENNIAL VEGETABLES AND HERBS

Most vegetables and herbs are treated as annuals and seed is sown each year, either in one sowing or in successive batches to give a long period of cropping. Some are hardy enough to be sown *in situ* outdoors, others are half-hardy and require the benefit of heat to start them into growth before planting out; while a third group requires the benefit of a greenhouse or frame throughout their lives.

Common edible plants that fall in the above three categories include indoor and outdoor tomatoes, cucumbers, marrows, melons, lettuce, endive, chicory, celery, radish, cabbages, cauliflower, broccoli, brussel sprouts, peas, beans, beetroot, spinach, carrots, parsnips, turnips, leeks and a number of herbs.

Those vegetables and herbs that are perennial and can be increased vegetatively are given in the following alphabetical list. Details of the methods recommended for propagation will be found in the preceding chapters.

Artichoke—globe (Cynara scolymus). Remove rooted suckers in early winter or late spring.

Artichoke—Jerusalem (Helianthus tuberosus). Plant small tubers in spring.

Asparagus (Asparagus officinalis). Division in spring; seed in spring outdoors.

Balm (Melissa officinalis). Division in autumn or spring; seed in late spring outdoors.

Bay—sweet (Laurus nobilis). Semi-hardwood cuttings in late summer; layer in summer.

Burnet (Sanguisorbia minor). Division in mid-spring.

Chives (Allium schoenoprasum). Division in early autumn; seed in spring outdoors.

Garlic (Allium sativum). Division of bulbs into 'cloves' (scales) in early spring.

Horseradish (Cochlearia armoracia). Root cuttings in spring.

Hyssop (Hyssopus officinalis). Basal softwood cuttings in late spring in cold frame; seed in spring in cold frame.

Marjoram—pot (Origanum onites). Division in spring; basal softwood cuttings in late spring in cold frame; seed in early spring in heat (13°C (55°F)).

Mint (Mentha species). Division in spring or autumn.

Onion (Allium cepa). Plant 'sets' (small bulbs) in spring; seed in early spring or autumn outdoors.

Potato (Solanum tuberosum). Sprout tubers in frost-free place in late winter and plant whole or cut tubers (each with sprouts upwards) in spring outdoors.

Rhubarb (Rheum rhaponticum). Division in autumn or spring; seed in spring outdoors.

Rosemary (Rosmarinus officinalis). Semi-hardwood cuttings in summer in cold frame; hardwood cuttings in late autumn or spring outdoors.

Sage (Salvia officinalis). Softwood cuttings in early summer in cold frame; seed in late spring outdoors.

Savory—winter (Satureja montana). Division in spring; softwood cuttings in early summer in cold frame.

Seakale (Crambe maritima). 'Thongs' (root cuttings) 15 cm (6 in) long taken in autumn, kept in bundles in sand in cold frame over winter, and planted outdoors in spring.

Sorrel (Rumex acetosa). Division in spring; seed in spring outdoors.

Tarragon (Artemisia dracunculus). Division in spring; basal softwood cuttings in spring in heat (13°C (55°F)).

Thyme—common and *lemon (Thymus vulgaris, T. × citriodorus)*. Division in spring; layer in spring; softshoot cuttings in summer in cold frame; seed in spring in cold frame.

Index

115